Advances in Information Security

Volume 93

Series Editors
Sushil Jajodia, George Mason University, Fairfax, USA
Pierangela Samarati, Milano, Italy
Javier Lopez, Malaga, Spain
Jaideep Vaidya, East Brunswick, USA

The purpose of the *Advances in Information Security* book series is to establish the state of the art and set the course for future research in information security. The scope of this series includes not only all aspects of computer, network security, and cryptography, but related areas, such as fault tolerance and software assurance. The series serves as a central source of reference for information security research and developments. The series aims to publish thorough and cohesive overviews on specific topics in Information Security, as well as works that are larger in scope than survey articles and that will contain more detailed background information. The series also provides a single point of coverage of advanced and timely topics and a forum for topics that may not have reached a level of maturity to warrant a comprehensive textbook.

Dietmar P. F. Möller

Cybersecurity for Network and Information Security

Principles, Techniques and Applications

Dietmar P. F. Möller
Clausthal University of Technology
Clausthal-Zellerfeld, Germany

ISSN 1568-2633   ISSN 2512-2193 (electronic)
Advances in Information Security
ISBN 978-3-031-99789-1   ISBN 978-3-031-99790-7 (eBook)
https://doi.org/10.1007/978-3-031-99790-7

© The Editor(s) (if applicable) and The Author(s), under exclusive license to Springer Nature Switzerland AG 2026

This work is subject to copyright. All rights are solely and exclusively licensed by the Publisher, whether the whole or part of the material is concerned, specifically the rights of translation, reprinting, reuse of illustrations, recitation, broadcasting, reproduction on microfilms or in any other physical way, and transmission or information storage and retrieval, electronic adaptation, computer software, or by similar or dissimilar methodology now known or hereafter developed.

The use of general descriptive names, registered names, trademarks, service marks, etc. in this publication does not imply, even in the absence of a specific statement, that such names are exempt from the relevant protective laws and regulations and therefore free for general use.

The publisher, the authors and the editors are safe to assume that the advice and information in this book are believed to be true and accurate at the date of publication. Neither the publisher nor the authors or the editors give a warranty, expressed or implied, with respect to the material contained herein or for any errors or omissions that may have been made. The publisher remains neutral with regard to jurisdictional claims in published maps and institutional affiliations.

This Springer imprint is published by the registered company Springer Nature Switzerland AG
The registered company address is: Gewerbestrasse 11, 6330 Cham, Switzerland

If disposing of this product, please recycle the paper.

Foreword 1

The worldwide rapid change in technological innovations is reviving all industrial, public, and private sectors and boosting their economy and sustainability. Such a rapid change, however, increases cybersecurity risks to products, processes, strategies, services, and others that must be defended. To defend against cyberattacks, industrial sectors and organizations must be cybersecure. Given this fact, this book provides a systematic overview of core research areas and measures, methods, and technologies with a focus on cybersecurity in network and information systems.

In this context, fundamental cybersecurity goes beyond mere protection; it encompasses cybersecurity situation awareness, risk assessment, risk management, maturity level models, risk mitigation, and other essential topics. Due to the sophistication of cybersecurity risks, it is also necessary to understand international cybersecurity standards like NIST CSF 2.0, ISO/IEC 27 K, MITRE ATTACK, CIS CSC, and the NIS2 Directive to defend against unprecedented global cybersecurity risks.

The well-written chapters of this book showcase the author's academic rigor and professionalism. It provides an important reading on cybersecurity measures, offering new perspectives and documenting important methods and progress in cybersecurity-related analysis and research.

Knowing Prof. Dr. Dietmar P.F. Möller well for over 25 years and witnessing firsthand his scholarly research and active teaching methods in transatlantic courses, I do not doubt that this book will be well received. It provides the most helpful learning concepts to students, academic colleagues, and industry experts seeking to gain more knowledge about advanced methodologies in cybersecurity in network and information systems.

As an adjunct professor in the Electrical and Computer Engineering Department of the University of Nebraska, Dr. Möller's research and expertise in cybersecurity have been a valuable addition to our student learning. I can say without reservation that this book, and, more specifically, the methods it describes, will change fundamental ideas for a better understanding of innovations and opportunities required in cybersecurity today. As a result of Dr. Möller's efforts, I believe there will be many grateful readers for his research and timely publication of this book, which provides a broader perspective of cybersecurity for readers.

Hamid Vakilzadian
Department of Electrical and Computer Engineering
University of Nebraska Lincoln,
Lincoln, NE, USA

Foreword 2

Cybersecurity has emerged as one of the most important needs on the research front today as evidenced by its nearly daily appearance in cyberattacks with loss of millions of files containing personally identifying information, ransom attacks that affect industrial OT systems and cost an organization millions of ransom, and many other cyberattack types. Industry has developed and implemented many new technologies in attempts to improve the cybersecurity protecting their mission critical IT and OT infrastructure, but cyber breaches still despite and the resulting penetrations cause reputational, economic, and physical harm. Examples include industrial control system cybersecurity, cybersecure the Internet of Things and the Industrial Internet of Things, and other, like breaches in all other public and private areas. It is these latter concerns that Dr. Dietmar P.F. Möller has chosen to address because research topics involving cybersecurity are becoming increasingly important in all organizations and societies.

Rapid advances being made in the connectedness in manufacturing including machine-to-machine communication, and the collection/storage of production information by the respective system, over-the-air firmware updates, cloud computing and cloud services, digital transformation in public and private sectors, including the introduction of artificial intelligence into the operation of industrial production, and others. However, these unprecedented technological innovations are also known and applied by today's cyberattackers. Therefore, the case for enhanced proofs of correctness, stronger verification and validation techniques, formal models, and code proofs must be stronger today than in the past. A treatment of these subjects through case studies and narrative is needed, and I compliment Dr. Dietmar P.F. Möller for taking up the challenge of writing this book and giving it a practical perspective with detailed insight into real-world applications. I am quite comfortable that this book will find its way into academic settings as well as industrial R&D organizations and that it will promote the kind of thoughtful dialogue necessary to avoid that vulnerable systems become deployed.

Gregory A. Harris
Department Chair Industrial and Systems Engineering
Samuel Ginn College of Engineering
Auburn, AL, USA

Interdisciplinary Center for Advanced Manufacturing Systems (ICAMS)
Auburn, AL, USA

Auburn University Rural Partnership Institute (AURPI)
Auburn, AL, USA

National Center for Additive Manufacturing Excellence (NCAME)
Auburn, AL, USA

Preface

Cyber threat risk incidents continue to advance in their sophistication and prevalence. Therefore, a high cybersecurity maturity level protection of organizations' sensitive data and business assets with regard to processing, storage, or exchange with third parties is required. With it, the protection of sensitive data and business assets in respect of processing activities needs to ensure the free and undisturbed flow of data. Given that fact, coherent data protection frameworks are required backed by strong cybersecurity enforcement, creating the trust that enables the digital economy to growth internationally. This global phenomenon is capturing attention in every organization and spurring major investment. However, it is not only a single objective; it is multifaceted depending on the organization's goals and its digital maturity level. So, the goal of this book is to provide a comprehensive, in-depth, and state-of-the-art overview of trends, principles, methods, technologies, applications, and best practices in cybersecurity for network and information security. Thus, the book provides an ideal framework to understand the complexity of cybersecurity issues and comprehensive cybersecurity in network and information security. For this reason, some choices were made in selecting the material for this book. For this purpose, the framework conditions of the European Union (EU) Network and Information Security Directive (NIS2) is introduced in detail and international cybersecurity frameworks like NIST CSF, MITRE ATTACK, CIS CSC, and ISO/IEC 27 K are taken into consideration.

Against this background, the book provides the reader with a format to assimilate the associated topic-related requirements. Without such a reference, the practitioner is left to ponder the plethora of terms, techniques, standards, and practices independently, which often lack cohesion, particularly in nomenclature and emphasis. Therefore, this book intends to cover all aspects in network and information security and provide a framework for the consideration of the many objectives associated with cybersecurity. These subjects are discussed with regard to trends, principles, methods, techniques, applications, and best practices to help the reader master the material.

This book can serve as a reference book or a textbook for college courses on network and information security and its cybersecurity issues and can be offered in computer science, electrical and computer engineering, information technology and information systems, applied mathematics and operations research, business informatics, and management departments. The contents of the book are also very useful to scientists interested in cybersecurity in network and information security, as well as engineers who apply the principles described in the book for their development work.

The use of the book can be a primary text in a course in various ways. It contains more materials covered in detail in a quarter-long (30-h) or semester-long (45-h) course. Instructors may select their own topics and add their own case studies applying the book material. The book is also suitable for self-study for engineers, scientists, computer scientists, and CIOs for on-the-job training, as well as study in graduate schools, and as a reference for practitioners and scientists in cybersecurity in network and information security.

However, a textbook cannot describe all innovative aspects of cybersecurity in network and information security. For this reason, reference is made to specific supplemental material, such as references, other textbooks, case studies, and Internet-based information that address to topics selected for the book. The book contains four chapters that can be read independently or consecutively.

Chapter 1 Digitalization and Cybersecurity: With digitalization organizations can store, process, and visualize data in new ways to speed up with decision-making, enhance new products, productivity, services, and others. This changes, for example, business models and operations as well as market requirements. Therefore, Chap. 1 introduces in Sect. 1.1 evolving features of digital transformation with their unprecedented impact on organizations to enhance their business success. Section 1.2 focusses on challenges and obstacles in digital transformation, while Sect. 1.3 introduces cybersecurity, which aims to reduce cyber threat risk incidents, for example, protecting against unauthorized exploitation of systems, networks, and technologies. From the practical perspective, Sect. 1.4 provides guidance for establishing cybersecure Operational Technology (OT), while Sect. 1.5 introduces the well-known CIA triad, a globally accepted model in information systems cybersecurity. Finally, Sect. 1.6 describes why actionable knowledge in cybersecurity is paramount and should be situational aware and sustainable, which also considers modern AI applications. Section 1.7 contains comprehensive questions, followed by references for further reading.

Chapter 2 Network and Information Security—NIS2: The practice protecting network and information systems from cyber threat risk incidents is part of cybersecurity efforts. NIS2 sets out rules for a regulatory directive and lays down cooperation mechanisms among relevant authorities in each EU Member State. Given that fact, Chap. 2 introduces in Sect. 2.1 the background that applies to the NIS2 Directive. Section 2.2 focusses on NIS2 Chapter I General Provision, and Sect. 2.3 on NIS2 Chapter II Coordinated Cybersecurity Frameworks. Section 2.4 refers to NIS2 Chapter III Cooperation at Union and International Level, and Sect. 2.5

introduces in detail NIS2 Chapter IV Cybersecurity Risk-Management Measures and Reporting Obligations and gives detailed hints about methodological, practical, and application-oriented solutions. Section 2.6 focusses on NIS2 Chapter V Jurisdiction and Registration, while Sect. 2.7 refers to NIS2 Chapter VI Information Sharing, and Sect. 2.8 focusses on NIS2 Chapter VII Supervision and Enforcement. Finally, Sect. 2.9 introduces NIS2 Chapter VIII Delegated and Implementing Acts, while Sect. 2.10 focusses on NIS2 Chapter IX Final Provisions. Section 2.11 contains comprehensive questions, followed by references for further reading.

Chapter 3 Cybersecurity Practices for NIS2 Measures: Cybersecurity is required to protect organizations crucial and critical data and assets against phishing schemes, ransomware attacks, identity theft, data breaches, and others. Thus, cybersecurity encompasses methods, technologies, processes, and practices to defend network and information systems against cyber threat risk incidents. Against this background, Chap. 3 introduces in Sect. 3.1 risk-management and assessment with regard to the effectiveness of risk-management measures. Section 3.2 focusses on cybersecurity frameworks such as the international cybersecurity standards NIST CSF 2.0, ISO/IEC 27 K, MITRE ATT & CK, and others. Thereafter, Sect. 3.3 introduces cybersecurity maturity based on Cybersecurity Maturity Model (CMM) and Cybersecurity Maturity Model Improvement (CMMI). Section 3.4 contains comprehensive questions, followed by references.

Chapter 4, Application Domain Network and Information Security (NIS2): The Network and Information Security (NIS2) sets out distinct measures aimed on achieving a higher cybersecurity maturity level across the EU Member States. Hence, Chap. 4 introduces in Sect. 4.1 aims and scope of NIS2 from an application-oriented perspective, while Sect. 4.2 refers to the compliance and regulatory pressure that results from NIS2. Section 4.3 focusses on liability issues referring to NIS2, while Sect. 4.4 gives essential practical measures usable for organizations with regard to the requirements in NIS2 Article 21.2 that must be fulfilled. Section 4.5 focusses on how organizations should prepare for NIS2 requirements and Sect. 4.6 gives important practical examples for realizing business continuity planning with regard to NIS2. The focus of Sect. 4.7 is on the elementary requirements to build up an emergency communication plan that fits into NIS2. Section 4.8 contains comprehensive questions, followed by references.

Beside the methodological, technical, and application-oriented content, all chapters in the book contain chapter-specific comprehensive questions to enable readers to determine if they have gained the required knowledge. This also allows to identify possible gaps in knowledge to be closed. Moreover, all chapters include references that also act as suggestions for further reading to enable readers to be effective and efficient in cybersecurity resilience. Hence, the book offers the opportunity to engage in the exciting cybersecurity domain applying it. This book provides knowledge and measures to cybersecurity in network and information systems, it is neither intended as legal advice nor should anybody consider it as such, because the described requirements of directives and frameworks may change in between.

I would like to thank Susan Lagerstrom Fife, Springer Publ., for her help with the organizational procedures between the author and the publishing house. Furthermore,

I sincerely thank all authors who have published material about cybersecurity, network and information security, and/or digitalization and digital transformation, which contribute to this book through citation.

Most notably, I am deeply grateful to my wife Angelika and appreciate her encouragement, forbearance, patience, and understanding without which this book would never have been written.

Clausthal-Zellerfeld, Germany   Dietmar P. F. Möller

Competing Interests   The author has no competing interests to declare that are relevant to the content of this manuscript.

Contents

1	**Digitalization and Cybersecurity**.............................	1
	1.1 Digitalization in Digital Transformation	1
	1.2 Challenges and Obstacles in Digital Transformation............	11
	1.3 Cybersecurity...	14
	1.4 Operational Technology Security.............................	34
	1.5 CIA Triad..	42
	1.6 Cybersecurity Is Paramount..................................	48
	1.7 Exercises	66
	References..	69
2	**Network and Information Security (NIS2)**......................	75
	2.1 Introduction to Network and Information Security (NIS2).......	75
	2.2 Chapter I General Provisions................................	82
	2.3 Chapter II Coordinated Cybersecurity Frameworks	84
	2.4 Chapter III Cooperation at EU and International Level.........	87
	2.5 Chapter IV Cybersecurity Risk-Management Measures and Reporting Obligations.................................	91
	2.6 Chapter V Jurisdiction and Registration	137
	2.7 Chapter VI Information Sharing	139
	2.8 Chapter VII Supervision and Enforcement	141
	2.9 Chapter VIII Delegated and Implementing Acts	144
	2.10 Chapter IX Final Provisions................................	145
	2.11 Exercises	146
	References..	147
3	**Cybersecurity Practices for NIS2 Measures**.....................	151
	3.1 Risk-Management and Assessment of Effectiveness of Risk-Management Measures.............................	151
	3.2 Cybersecurity Frameworks and Criteria	170
	3.3 Cybersecurity Maturity Models.............................	196

	3.4	Cybersecurity Maturity Assessment	209
	3.5	Exercises	211
		References	212
4	**Application Domain Network and Information Security (NIS2)**		**217**
	4.1	Network and Information Security (NIS2)	217
	4.2	Compliance and Regulatory Pressure	221
	4.3	Liability	224
	4.4	NIS2 Article 21.2	226
	4.5	Preparing for NIS2	234
	4.6	Business Continuity Plan	236
	4.7	Emergency Communication Plan	246
	4.8	Exercises	250
		References	251

Glossary 253

Index 267

Chapter 1
Digitalization and Cybersecurity

1.1 Digitalization in Digital Transformation

Today's technology landscape changing faster than ever before, whereby the evolution of digitalization become mainstream to the digital world. Digitalization enable unprecedented opportunities to all sectors of industry and society today, based on digital data as new raw material. Digital data is information represented as a string of discrete symbols, each of which can take one of only a finite number of values from some alphabet, such as letters or digits. Digital data are generated, collected, accessed, analyzed, processed, extracted, managed and stored. Thus, digitalization is the driver for digital data usage, whereby the advantages processing digital data are incredible because they can be analyzed, distributed, manipulated, reproduce stored and others by using digital data processing systems like computers. Computer use the machine-readable form of digital data that make them faster to process and search, for example, specific content in digital data. Over and beyond, compression algorithms applied to digital data significantly reduce storage requirement, resulting in a significant impact to the specific needs in digital transformation, and others. The term digital transformation itself broadly defines the ability, for example, of all kinds of industrial sectors, to adopt, apply and use digital technological capabilities to fulfill their digital business. At the intersection of business and technology, digital transformation guiding industrial sectors towards a more agile, efficient, as well as customer-centric direction, leading to fundamental changes in how industrial organizations operates their operational business and delivers added value to their customers. However, digital transformation is a complex process that involves integrating new digital technology into all areas of organizations to thrive operational efficiency, flexibility, innovation and growth. With it, digital transformation requires industrial organizations commitment to direct and indirect effects of the usage of digital technologies and techniques, processes and culture on organizational and economic conditions, new products and services, to bring toward the digitalization goal together. The direct and indirect effects of technology, processes and culture are:

- **Technology:** Improve efficiency and support digitalization regardless of whether an on-prem, a cloud, or a hybrid environment is used. In general technologies in digitalization and digital transformation are artificial intelligence, big data and analytics, cloud computing and services, internet of things and industrial internet of things, robotics process automation and others, which enable industrial business to automate repetitive tasks and improve efficiency. In this context, platforms with established community support in open source-based digital tools are preferred.
- **Processes:** In general, processes execute the necessary activities to achieve their digital and corporate objectives by digitally managing data, to swiftly deploying changes throughout a digital environment. Deploying changes is governed by change management processes that need to support a digitalized approach.
- **Collaboration Culture:** Break silos and working across functional teams to swiftly solve complex problems, streamline processes and build digitalized flows connecting different platforms and systems.

In this regard, digitalization and digital transformation is concerned with changes that new digital technologies bring into industry's business models, production, services, collaboration or organizational structures. Moreover, this is the most pervasive challenge for all sectors of industry and society of the last and decades coming. In this context, digitalization and digital transformation involves a sophisticated variety of advanced and intelligent technologies and skills to understand, develop, and dominate them to make business, governmental, industrial, and society processes more innovative, intelligent and efficient. This fundamentally changes the ways to operate and deliver value to customers. However, the idea behind is to use digital technology not just to replicate existing process in a digital form, but use digital technology transform that process into something intelligent, where anything is connected with everything at any time and accessible, controllable and finally significantly designable in an advanced manner. Thus, beside digital technologies and processes, digitalization and digital transformation needs skilled employees and executives in the collaboration culture, in order to reveal its transformative power due to resultant innovative business models and altered consumers expectations, which has enormous pressure on traditional business models [1–3].

Against this background, it should be noted that the transformation of industry through technological innovations occurred in several technological waves. In this context, todays fourth industrial transformation optimizes the computerization of the third industrial transformation through a manifold of digital innovations such as wireless connectivity, wireless infrastructure devices and networks, intelligent algorithms, virtualization, cloud computing and cloud services and others, which result in the digital transformation paradigm [4, 5]. With it, digital transformation refers to the advent of digital innovations that comprises implanting new combinations of digital and physical components to generate novel outcomes [6]. In this regard, the confluence of technologies is fundamentally changing how industrial organizations can operate their business to become competitive to survive and thrive in their market, which is constantly evolving and leading to groundbreaking digital- and emerging technologies such as:

- **Additive Manufacturing:** Process used to fabricate a physical object by adding material rather subtracting it, under which production processes such as rapid prototyping, rapid tooling or mass customization be assumed. Additive Manufacturing based on a three-dimensional digital model. With this model, the manufactured objects typically made by laying down and bonding a large number of successive thin layers of materials. The tools used to layer the material in this procedure are operated digitally controlled. Relevant processes and applications in Additive Manufacturing are aerospace industry, manufacturing industry, medical and dental applications and others. Aerospace industry employs Additive Manufacturing because of the possibility of manufacturing lighter structures to reduce weight. A number of advantages for organizations arise for additive manufacturing such as the Additive Manufacturing-as-a-Service (AMaaS) principle, where organizations can operate as contractors on demand, creating products for their businesses that employ them while distributing costs for software, equipment, maintenance, and repairs, whereby
  - AMaaS can create efficiency and boosts capacity utilization,
  - AMaas can create higher quality products,
  - AMaaS can make changes to any production line to reduce errors and improve processes.
  - And others.

- **Artificial Intelligence (AI):** Theory and development of computer systems to make them able to perform tasks normally requiring human intelligence. Specific AI applications include decision-making, expert systems, machine vision, Natural Language Processing (NLP), speech recognition, visual perception, and others. NLP is a Machine Learning (ML) technology that enables computer systems to interpret, manipulate, and comprehend human language. Furthermore, Artificial Intelligence enable tailored applications across major industries, from healthcare and finance to manufacturing and retail, and others, because the transformative impact of Artificial Intelligence is reshaping traditional business paradigms. By leveraging Artificial Intelligence algorithms in form of predictive analytics, and advanced automation, businesses harness the power of Artificial Intelligence to streamline operations, optimize resource utilization, gain insights into consumer behavior and market trends, and in turn is one of the primary drivers of Machine Learning (ML) use in industrial, public and private organizations. A number of advantages for organizations arise when using Artificial Intelligence:
  - Save Time and Money by automating and optimizing routine processes and tasks,
  - Increase Business Decisions based on outputs from cognitive technologies.
  - Increase Productivity and Operational Efficiencies,
  - Avoid Mistakes and Human Error,
  - And others.

- Artificial Intelligence also plays a crucial role in early detection of cyber threat risk incidents. Analyzing large volumes of data it can identify anomalies and suspicious behavior in network and information systems, enabling rapid responses and effective countermeasures.
- **Big Data and Analytics:** Describes the process of uncovering trends, patterns, and correlations in large amounts of raw data (big data) to enable data-informed decisions in digital transformation. Big Data is an information asset, characterized by high volume, velocity and variety to describe the characteristics of information, achieved by specific technological and analytical methodological requirements. The main methods of Big Data and Analytics are:
  - **Descriptive Analytics:** Analyzing historical data so that patterns and trends become apparent, which enable adjust strategies and correct problems. Descriptive analytics is applied, for example, in Business Intelligence Application (BIA). Furthermore, a descriptive model also is an equation chosen to fit experimental or observational data,
  - **Diagnostic Analytics:** Aims to understand factors that lead to past events and outcomes and why these trends occur. Process similar to regression analysis that identifies the variables that impact a given scenario,
  - **Predictive Analytics:** *Using data to determine the causes of trends and correlations between variables.* Aims to predict upcoming events and future trends, enabling for more accurate forecasts and improved planning processes. This always relies on statistical algorithms for informed peeks how business conditions will evolve in the future. Artificial Intelligence (AI) and Machine Learning (ML) improved the accuracy of predictive data modeling scenario,
  - **Prescriptive Analytics:** *Use data to predict future trends and events, to make choices providing suggestions and recommendations based on the insights gleaned from* data analysis, AI, ML, statistical models and historical trends to find patterns that might predict future behavior. *Hence it can provide decisions for example for a given business scenario to choose best action to be taken, ensuring that each choice is best fit for most favorable outcome,*
  - **Cognitive-Analytics:** Intelligent analytics based on multiple analytical techniques to analyze large data sets to give structure to unstructured data. It utilizes advanced AI such as Natural Language Processing (NLP) and Deep Learning (DL), to extract insights and make predictions from this information, for example, in automated-driving cars that uses ML to navigate and make decisions on the road situation.
- Big Data and Analytics are a significant component of digital transformation within the Industry 4.0 (I4.0) paradigm, as it provides valuable insights for the purpose of intelligent or smart manufacturing management, which requires that data be processed with advanced tools and technologies in order to provide relevant information [7, 8], such as large-scale device data, production life-cycle data, organizations operational data, manufacturing value chain data, external cooperation data, and others. In this regard, data modeling and analytics are inte-

gral part of almost any data driven decision-making in the digital transformation era, where Big Data technologies are shifting from data collection to data analytics and data outcome [9]. However, big data represent data that is just too large to be managed by traditional databases and processing tools. The problem of big data lies in the diverse data structures and that they are difficult to analyze or incorporate in a traditional structural database. However, Big Data and Analytics is important because it enable organizations use colossal amounts of data in multiple formats from multiple sources to identify opportunities and risks, helping organizations swiftly move and improve their bottom lines. But not only collection of data increased like an avalanche but also the possibilities of their evaluation. Hence, it is no wonder that at the same time as the advent of Big Data rises, the corresponding advances in data processing increase. Nevertheless, analyze such gigantic amount of data no longer be analyzed without advanced computational aids that even the usual methods of digital data processing are no longer sufficient. Decisive for this trend are advances in the fields of computer science and applied mathematics [10]. A number of advantages for organizations arise for Big Data and Analytics such as:

- **Data Analytics:** Guide business decisions and minimize financial losses, and enable organizations to understand risks and take preventive measures. Organizations can use data analytics to diagnose causes of past data events by processing and visualizing relevant data,
- **Predictive Analytics:** Suggest what could happen in response to changes to organizations business, and prescriptive analytics can indicate how organizations business should react to these changes. Thus, organizations improve operational efficiency through data analytics,
- And others.

• **Cloud-Computing and Cloud-Service:** Full and highly scalable deployment of compute, storage and application resources, accessible from anywhere, used by multiple industrial sectors in the era of digital transformation. Term used for anything that involves delivering hosted services over the Internet. The main technical under-pinning's of cloud computing infrastructures and services include virtualization, which is the creation of a virtual rather than physical version of an entity, such as network resources, server, storage device, and other, service-oriented software, grid computing technologies, management of large facilities, and power efficiency. User's purchase such services in the form of:

- **Infrastructure-as-a-Service (IaaS):** On-demand availability of highly scalable computing resources as services over the internet,
- **Platform-as-a-Service (PaaS):** Cloud computing service model that offers a flexible, scalable cloud platform to develop, deploy, run, and manage apps,
- **Software-as-a-Service (SaaS):** Allows users to connect to and use cloud-based apps over the Internet,
- and others,

- and sell value-added services (such as utility services) to users. Therefore, a number of business advantages for organizations arise using cloud computing:
  - **Predictable Costs:** Cloud services are typically paid on a monthly basis or based on use with little or no upfront costs.
  - **Reduced Total Cost:** Benefits of cloud services derived from economies of scale that a service provider can achieve.
  - **Access to the Best Available Technology:** Cloud services enable organizations to benefit from the best technologies, without making any initial costs.
  - And others.
- **Digital Twin:** Emerging and vital technology in digital transformation and an intelligent upgrade, driven by data and models, to perform monitoring, simulation, prediction, optimization, and other specific tasks. Digital Twin modeling is a core function for accurate portrayal of physical entities, which enables to deliver functional services and satisfy application specific requirements. Hence, a Digital Twin represents a real object of the physical world in digitalized form, using data and algorithms, describing properties and behavior of real-world objects under certain conditions that be connected to the real physical world in real-time via smart sensors. In the course of coupling real world data, such as environmental conditions or machine positions, a Digital Twin enable complex analyzes and simulations of development and production processes, which are the basis for digitalization of intelligent manufacturing processes in I4.0. The advantages of Industry 4.0 include improved productivity and efficiency, greater flexibility and agility, and increased profitability. In Industry 4.0, a Digital Twin accompany the entire development, production and operating cycle of a product or service. Processes can be planned, optimized and adjusted through virtual simulation models. Furthermore, Digital Twin can also be used simulating the manufacturing line before the actual implementation of the real manufacturing shop floor.
- **Industrial Internet of Things (IIoT):** Network of connected intelligent objects like smart sensors, actuators and other devices, such as Radio Frequency Identification tags (RFID) that monitor, collect, exchange and analyze data, to enhance manufacturing and industrial processes [11]. Since misuse of IIoT can have catastrophic consequences, creating high risk and potentially life-threatening situations. Thus, cybersecurity of the IIoT is an essential topic to defend malicious misuse of it, because cybersecurity in IIoT systems is crucial for their wide adoption. Without cybersecurity in communication and cyber threat risk incident detection and prevention techniques, it can become very difficult to effectively use IIoT in smart, connected systems in the industrial settings of digital transformation. A number of business advantages for organizations arise from IIoT like:
  - Enabling manufacturing organizations to automate their processes and improve their productivity and uptimes,
  - Improve process and operational efficiencies and product quality,

- Accelerate innovation and create end-to-end operational visibility,
- Enable manufacturing organizations to save costs and boost revenue down the line,
- And others.

- **Mobile Technology:** Important part of digital transformation that provide seamless interaction with customer at all touch points with business organizations through growth of a mobile digital business and connected worker platforms, which make it easy for frontline workers to be connected. Enable to integrate because the solution is on devices that workers are already familiar with, e.g. smartphones and tablets. Their expertise and experience transfer over to the job. Even for a multigenerational workforce, mobile solutions can be easy for workers to adapt to when it comes to operational procedures, daily checklists, processes and more. Mobile solutions are highly deployable and cause little friction when introduced to frontline workers, because the global digital economy work are mobility and digital collaboration with regard to the respective organizational cultural transformation. Thus, the use of mobile technologies in business and the current level of integration between digital technologies entirely caused by the needs of the organizations and focused on optimal business processes management. A number of business advantages for organizations arise from mobile technology in digital transformation:

  - **Mobile Solutions:** Highly deployable and cause little friction when introduced to frontline workers,
  - **Outcome:** Capturing data at the most important point in the improvement cycle with the frontline worker at the point where work occurs,
  - **Organization:** Can ultimately propel own digital transformation forward with mobile technology for greater productivity and being more profitable than normal.

- And others [12].

Moreover, digitalization and the digital transformation paradigm require adopting the business-related practices of digital interactive services like Business-to-Business (B2B) and Business-to-Customer (B2C), to build personal relationships with customers. B2B services primarily focus on building long-term relationships to boost business growth, based on personal emotional relationship on understanding all the parties involved such that their roles in the business are fortified enough to last long. On the other hand, B2C service efforts are targeted directly at customers and businesses try to forge an emotional bond with their present and future customers. The focus is largely on pushing sales. Furthermore, the business-related Business Process Management approach uses various methods to discover, model, analyze, measure, improve and optimize business processes, which have a positive impact on business application and data within digital ecosystems. This enables organizations to control new and/or old technologies, and build automated processes around them that consistently grow their business within the digitalization and digital transformation era. In this context, Business-Process Management

(BPM) is a systemic approach to capture, document, execute, shape, measure, monitor and steering automatic and non-automatic processes, to reach coordination and sustainable industrial organizations targets, improving corporate performance, by optimizing and managing business processes of the respective organization [13]. Given that fact, digitalization, digital transformation and digital ecosystems interact with each other, combining digitalization and the individual digital transformation approach with ecological thinking of the digital ecosystem strategy. The strategic goal is to clarify what an industrial organization has to work with, to guarantee they have the right digital tools supporting their goals and ensure they are being as efficient and effective as possible, to deliver and exceed the organizations expectations with a robust digital mindset, backed by innovation through digital transformation. However, digitalization and digital transformation are not unified entities, they much more refer to core characteristic subject topics such as:

- **Business Transformation:** Focus on certain aspects of the industrial organization's business model with fundamentally changing systems, processes, technology and people across a whole organizations business or business unit and third parties, which also require specific programs for recruiting, recognition, retention and referrals, to achieve measurable improvements in efficiency, effectiveness and stakeholder satisfaction to provide value in a particular industrial sector. In essence, businesses are utilizing digital transformation to change traditional business models,
- **Cultural Transformation:** Culture is often perceived as a valuable strategic asset that has the potential to support business transformation and the exploitation of digital technologies [14, 15]. However, organizational culture can also be the source of inertia that prevents change, like individualization that mostly use different approaches instead of getting different people using the same approach and accept resulting big changes. Thus, an organizational and cultural transformation is essential to fully exploit the benefits from the establishment of the digitalization and digital transformation mindset and change of culture is essential for successful the digital transformation [16, 17] due to the intrinsic innovation speed, disruption and competitive dynamics of digital transformation, which means acquire all digital transformation skills needed to accelerate transforming the business.
- **Domain Transformation:** Occur when an organization effectively transitions into another domain, e.g., developing a product to automate processes and decides to market this product to other organizations, thereby establishing a new revenue stream. However, a domain transformation may not receive as much attention in the digital transformation sector, but it can be a hugely successful form of transformation,
- **Process Transformation:** Addresses transforming business key issues like processes, technologies, services and others, including operational performance, customer relations and talent retention. It relies on robust and innovative digital solutions, with enterprise platform and AI at the forefront, but also to become a more efficient organization, get more satisfied customers, and more fulfilled, committed and loyal employees.

Against this background, digitalization and digital transformation transcends traditional features and roles in product design and development, product marketing and sales, customer-oriented services and others. With it, digitalization and digital transformation begins and ends with enabling intelligent digital technologies in their design and development strategy to think about and engage with customers from paper to spreadsheets to smart applications, to manage all business processes and values guided by a digital mindset. Therefore, digitalization and digital transformation changing the way business can perform in different shapes to create entirely new business asset, models, and capabilities. However, this can only be successful if there is a well-founded strategy and leadership behind, because transformational changes are required to implement the digital transformation requirements. The economic strategy behind is based on different responsibilities and roles for the economic actors as well as differing notions of resource mobilization and allocation, as well as leadership. Leadership in economy lies in the belief that identifiable, constant realities exist within the leadership reality that, when pursued and applied in a systematic way, will result in the ability to lead in ways that enable achieve desired outcome. Finally, economic strategies can employ more or less direct means to influence the mobilization and disposition of resources and the incomes they generate. However, digital transformation also causes a digital disruption. The term disruption has several connotations. It refers to a specific process that explain how entrants in transformation can successfully compete with the incumbents. The competitive relationship between incumbents and entrants, and the specific means through which the latter enter the market, which are key boundary conditions for digital disruption of digital transformation, rarely exhibited by examples of digital disruption [18, 19].

As reported in [20] the impact of digital transformation will be to organizations' value proposition, the customer segments they can identify and serve, the way organizations reach their customers, and the resources they use. Thus, many different changes are expected across different business model elements. Throughout the variety of changes there is a group that benefits the most; customers. Hence, the research outcome in [20] stated the value proposition of customers' is their value for money that increases a lot. Products and services are expected to increase in customization, performance, accessibility and convenience whilst prices are expected to decrease a little. Therefore, advanced competencies must be made accessible, available and aware, the 3A principle, as essential resource and scope in digital transformation. This include beside awareness of key competence in digital transformation understanding the potential of emerging technologies, which raise the question: *What is the digital and new technology used capable of, and how it can adapt to the required business and production processes to make the most out of technology investments and transformation?*

Based on stringent decisions answering this question, digital transformation enables the application of smart technologies to build new business models, processes, software and systems that results in profitable revenue, greater competitive advantage, and higher efficiency. Businesses achieve this by transforming processes and business models, empowering workforce efficiency and innovation in design, development, production, and services through personalizing customer experiences. As with any emerging technology, however, there are significant challenges [21] to serve better its principal business and stakeholders, customers, business partners, supply chains, and others. Digital transformation also leads to an increased agility within IT and engineering teams enabling them executing projects and processes in a more sophisticated, efficient and faster way. To achieve these goals, leading is essential to create the required digitalization environment for digital transformation. It also calls for proactive measures to avoid obstacles hindering or constraining digitization in the development stage [22]. Hence, digital transformation enables digital and emerging technologies to sustainable support organizations business to achieve the following features:

- Speed up with development and design of products to minimize time-to-market with new products and services,
- Increase employee productivity in all areas as part of Continuous Improvement Process (CIP). In general CIP is the ongoing process of analyzing performance, identifying opportunities, and making incremental changes to processes, products, and personnel.,
- Increase sustainable responsiveness to customer requests,
- Gain more insight into individual customers requirements to better anticipate and personalize products and services,
- Improve sustainable customer service, especially in providing more intuitive and more engaging customer experiences.

This sophisticated variety of advanced and emerging technologies need to understand, develop, and dominate the digital transformation narrative, to transform business processes sustainable in the context of digital, innovative, efficient and effective processes, as much as possible. Thus, deep competences in digital systems and network processes as well as deep knowledge in digital and emerging technologies are essential to dominate the digital transformation. To gain their successful usage, a clear vision and deep understanding of business models and production processes across the complete industrial organization is a must. Thus, key factors for success are:

- Leadership in digital transformation is achieved,
- Investing in skills and talent development is achieved,
- Clear responsibility in digital transformation is achieved,
- Clear communication strategy is achieved,
- Digital transformation tools and processes are set in place,
- Cybersecurity awareness and training is set in place,
- And others,

which all play key roles in success in industrial organizations digital transformation, spurring a growing need for it across industrial sectors. Given that fact, it may be impossible to predict how this transformation will look at the end of this evolutionary step. The process of rapid digitalization, started in the late twentieth century and underwent rapid acceleration in the first two decades or the twenty-first century. As described in [23] "organizations that figure out how to breath Big Data, how to harness the power of this new resource and extract its value by leveraging the cloud, AI, and IIoT, will be the next to climb out of the data lake and master the new digital land". This include sophisticated competences such as digital awareness, knowledge in digitization and emerging technologies, relevant and unavoidable to digital transformation.

1.2 Challenges and Obstacles in Digital Transformation

Digital transformation is the proactive process of using innovative digital technologies to enable new or modify actual business processes, culture, and customer experiences to meet the challenging business, market and technological requirements. Digital transformation results in changing traditional isolated activities/processes into connected, dynamic and rapidly evolving data flow driven activities/processes. Thus, it is a process of change that goes hand in hand with unprecedented innovation cycles enabled by the inherent dynamic development of digital technologies. At the same time, this paves the way for further and challenging technological innovations, such those seen in the development cycles of smartphones. The driving force behind is the change in customer experience and requirements, which can be served through innovative and emerging technological developments. However, one of the foremost challenges and obstacles in digital transformation is often the resistance to change within organizational culture. For example, one obstacle may be using siloed operational technological systems from different vendors and different integrations that are necessarily not designed to communicate to each other, which avoid data aggregation. Another obstacle may be missing knowledge in digital transformation processes and technologies, its tools and the learning curve within real organizations that often may require to embed external digitalization expertise. Thus, leadership as well as employees must be aligned with the digital transformation vision, fostering a mindset that embraces innovation and continuous learning. To become successful this require align employees proactively on the digital transformation plan before implementing required digital technologies to enhance process and technological knowledge. Unfortunately, often a notable lack of professionals with the specific expertise for digital transformation exists, presenting a notable talent challenge for organizations.

At the intersection of business and technology, digital transformation has emerged as roadmap guiding organizations toward an agile, efficient, and customer-centric attitude. Against this background, integrating digital technology and services into all areas of organizations drives operational efficiency, flexibility, innovation, growth, to thrive the future. It also requires a holistic shift in how organizations work and communicate. These challenges are compounding as organizations need, for example, moving workload to the cloud, or use generative AI and Large Language Models (LLMs) to adapt the requirements of digital transformation in regard to decision-making and their business operations to improve business outcome. LLM is a foundational model developed to understand, interpret and generate text using human language, by providing big datasets and finding patterns, grammatical structures and even cultural references in the datasets to generate text by using conversational language. LLMs are Deep Learning (DL) models that can be introduced as the brain model, being a chatbot, as implemented in OpenAI's ChatGPT (see Section "ChatGPT"). They work by analyzing vast amounts of data, which refers to the term large. DL is a type of ML, based on Artificial Neural Networks (ANN) in which multiple layers of processing are applied to extract higher level features from data.

Since the digital landscape evolves, organizations need to reconsider their current approach to cybersecurity. The reason for this is that cybersecurity has evolved as most critical and crucial issue to avoid cybercriminals cyber threat risk incidents. Given the foregoing fact moving workload to the cloud, or use Generative AI (GenAI), a type of AI that generate various types of content, including text, imagery, audio and synthetic data, and LLMs, which means the cyber threat risk incident surface is increasing. Therefore, organizations need to ensure they are protected against potential cyber threat risk incidents.

Cyber threat risk incidents can affect business's operations, damage its reputation, and leak sensitive business assets and data, such as customer records, trade secrets and others. Unfortunately, these cyber risks continue to evolve. Over and beyond, organizations are also concerned about cyber threat risk incidents associated with third-party code either generated by AI applications or from open source repositories, which requires cybersecurity criteria to efficient and successful planning of digital transformation strategic criteria, as shown in Table 1.1 [10].

Against this background, the main driver of the digital transformation challenge is technological based added value through methods such as:

- **Artificial Intelligence (AI):** Can be interpreted as the simulation of human intelligence processes by machines, especially computer systems,
- **Big Data and Analytics:** Refers to methods, tools, and applications used to collect, process, and derive insights from varied, high-volume, high-velocity data sets,
- **Cloud Computing and -Services:** Refers to the use of hosted services, such as data storage, servers, databases, networking, and software over the internet,
- **Connectivity:** Ability of a computer system, program, device, or component to connect to the internet, other systems, and others,
- **Cybersecurity:** Usage of technologies, processes, and controls to protect systems, networks, programs, devices and data from cyber threat risk incidents,
- **Interconnectedness:** Means having different parts or things or objects connected or related to each other,
- And others.

In this context, an emerging theme and frontier in digital transformation is the digital sovereignty. Digital sovereignty describes an organization's right and ability to control its own digital data, which means it enables an organizations business to effectively and compliantly handle its digital data, and staying in line with national/

1.2 Challenges and Obstacles in Digital Transformation

Table 1.1 Strategic criteria for successful digital transformation planning

Strategic criteria	Capability
Data	Key enabler of digital transformation success. Collecting data and perform analytics for better insights. This requires converting existing volumes of data into high valuable, actionable assets, and need to cybersecure data for digital transformation success
Digital Technologies	Requires multiple ongoing initiatives that involve investments in emerging technologies, new hard- and soft skills, and continuously updated cybersecure workplace culture to cybersecurity organizational restructuring
Digital Technology Risk	Requires continuously updated cyber threat risk incident awareness and risk management strategies by integrating emerged technologies with omnipresent connectivity to avoid unprecedented failures caused due to the manifold of potential cybersecurity risks, by using emerging technologies
Digital Transformation Project	Demonstration a compelling Return on Investment (ROI), vital for digital transformation success to support digital initiatives
Talent	Ability to compete in digital transformation depends on having the right talents at the right time and in the right place. This requires prioritizing finding for talent(s) needed, and establish a robust resource recruiting and management program to support planning for future skill requirements
Technology	To enable ongoing success requires embracing a platform-based operating model for digital transformation, considering all users, employees, customers, and partners in the context of technological enablers, driving customer experience, decision-making, process efficiency, scalability, performance, and comprehensive cybersecurity

international regulations. Ideally, organizations should aim for total digital sovereignty. As a business leader, the organization have total authority over all digital data and digital assets, and can take measures to keep it secure in the digital ecosystem, which involves considerations how digital data and digital assets are treated. Hence, over the last decade, digital sovereignty has become a key element in digital policy discourses, as described, for example, by European Union (EU) initiatives of the Digital Market Act (DMA) and the Digital Service Act (DSA). The DMA establishes a set of clearly defined criteria that apply across the whole EU to ensure that their online markets behave in a fair way and leave room for contestability. The DSA regulate online intermediaries and platforms to prevent illegal and harmful activities online. Both have two main goals:

- Create a safer digital space in which the fundamental rights of all users of digital services are protected,
- Establish a level playing field to foster innovation, growth, and competitiveness, both in the European Single Market and globally.

In this context, the DMA [24] establishes a set of clearly defined objective criteria to qualify, for example, large online platforms as so-called *gatekeepers* and ensures they behave in a fair way online and leave room for contestability, whereby obligations for *gatekeepers*, *do's* and *don'ts* must comply within their daily operations. In cybersecurity, a *gatekeeper* refers to processes or devices that manage access control to systems devices, websites and others. Hence a *gatekeeper* could be a login system that requires users to enter valid credentials before granting them access. Thus, a *gatekeeper* acts as a barrier, only allowing authorized users to access and view protected information.

In contrast, the DSA [25] aim to create a safer digital space where the fundamental rights of users are protected and to establish a level playing field for businesses. It includes a large category of online services, from simple websites to internet infrastructure services and online platforms. Moreover, the DSA includes specific rules for very large online platform and search engines. These are online platforms and intermediaries that have more than 45 million users per month in the EU. They must abide by the strictest obligations of the Act.

Users demand interoperability quality while finding it efficiently using of services through new data service intermediaries. This also includes transparency about the services offered. Users also want to be able to distribute data processing flexible across many providers and at the same time, which rely on robust processes to fall back on [26]. Thus, the next wave of digital transformation has to capture these challenges.

1.3 Cybersecurity

Digitalization and digital transformation are a worldwide emerging issue of major importance for all organizations in all sectors, reshaping business relationships, crucial processes in production, value creation and capabilities, improvements and others, and enables to connect everything with anything. Due to this interconnectedness organizations can swiftly become the focus of cybercriminals and become a victim of cyber threat risk incidents, which can threaten the reputation and existence of an organization if being successful. Therefore, cybersecurity awareness and improvements are an absolute must to survive and to thrive the organizations future. Cybersecurity, also termed as IT security, is the protection of connected entities, systems and applications from cyber threat risk incidents by combat them, regardless of whether these cyber threat risk incidents come from inside or outside of an attacked organization.

The term cybersecurity is used in a variety of contexts, from business application domains to mobile computing, and can be grouped into a number of categories as explained in [10]. Against this background, cybersecurity in general can be understood as practice to protect organizations against unauthorized access to sensitive data, business assets, and other valuable information, to avoid worst-case scenarios. In an industrial worst-case scenario, it often can happen that a production floor is paralyzed by a malicious cyber threat risk incident, so that production can come up to a complete standstill, while data may even be manipulated or changed. In addition, it also can happen that product specifications intended for a certain market are changed, because relevant parameters are manipulated by a malicious cyber threat risk incident, which may result in unusable or wrong products that end up on the market. Therefore, continuously improving cybersecurity is essential to trust, use, and benefit from technological innovation with their possibility of unprecedented connectivity and overall automation in the digital transformation age. This also include protecting the rights to privacy and to the protection of personal data, and information. Given that fact, cybersecurity is indispensable to the network connectivity and the global and open Internet that underpin the digital transformation of all industrial sectors and society. In this regard, directives and rules on cybersecurity are at the core of international cybersecurity directives, for example, the European Union (EU) Network and Information Security Directive (NIS2) [27], and the U.S. National Institute of Standards and Technology Cybersecurity Framework (NIST CSF) [28].

Analyzing the impact of digitalization and advancements in technologies in digital transformation, a major focus is on the topics: (i) knowledge about the interaction of digitalization and advanced technologies, (ii) knowledge about resulting intrinsic cybersecurity risks through the rapid advancements in digital technologies, shifting the cybersecurity risk incident landscape with their direct impact on cybersecurity. This potentially can have an adverse impact to organizations business objectives due to continuously conquer cyber risks effectively and sustainable. Thus, a cybersecurity strategy is required that aims to safeguard cybersecurity risk assessment as an integral part of organizations cybersecurity risk management, such as the EU Network and Information Security Directive (NIS2), and the U.S. National Institute of Standards and Technology Cybersecurity Framework (NIST CSF 2.0), in order to increase the level of cyber resilience in industrial, public and private organizations. Cyber resilience includes the essential management capabilities of information security, data protection and business continuity. In this context, information security refers to the overall practice of protecting personal data and the approaches to achieve that, while cybersecurity is a specific type of information security that refers to the ways that organizations protect digital information, such as networks, programs, devices, servers and other digital assets.

To comply with NIS2, organizations must take measures managing to mitigate and/or eliminate their cybersecurity risks. These measures include cyber threat risk incident management, supply chain security, network security, data security, access control, and others. Conducting a cybersecurity risk assessment, building an operational capability to prevent, deter, and/or respond to cyber threat risk incidents, which include

- **Identify Malicious Incidents:** Often result from cyber threat risk incidents that lead to undesired and negative operational business consequences of organizations.
- **Determine Cybersecurity Risk Levels:** Organizations mostly expose an unprecedented negative impact.
- **Create a Culture of Cybersecurity Risk Awareness:** Acquiring digital transformation skills to accelerate transforming business including a cybersecurity awareness culture.

In this regard, cyber risk is the likelihood that cyber threat risk incidents occur against organizations digital data, assets and crucial and critical entities, whereby the impact of the probable cyber risk occurrence result in real cybersecurity risk(s). Cyber threat risk incidents against organizations digital data and assets mostly based on vulnerabilities, whereby a cyber threat risk incident is exploited by a cyber-criminal by means of vulnerabilities.

Let's assume that a cybersecurity flaw exists and cybercriminal actors exploiting the vulnerability to compromise the targeted digital system, network, infrastructure resource, and others. In case the latest software patches that fix such a problem are applied, then the vulnerability can't be exploited, and the cyber threat risk incident has no chance being successfully executed. With it, the impact illustrates how much disruption an organization will face if a cyber threat risk incident occurs. Combining likelihood and impact, also known as consequence of cyber threat risk incidents, result in a residual risk, rating the risk tolerance score, e.g. between low, moderate, medium, high or unacceptable. Against this background, cybersecurity situational awareness, cybersecurity risk assessment and cybersecurity risk-management are important narratives of organizations and the essential part of the NIS2 Directive.

1.3.1 Cybersecurity Situational Awareness

Cybersecurity Situational Awareness (CSA) is the practice that foresighted to protect organizations against unauthorized access of business assets, digital data, processes, services, and systems, and its associated IT-risks, Cyber-risks, and Third-party risks and impacts on the business. CSA also stimulate the adequacy of IT-risks, Cyber-risks, and Third-party risks mitigation measures. Given that fact, CSA involves monitoring network and information system activity, identifying abnormal behavior, and recognizing potential vulnerabilities that cybercriminals could exploit. Furthermore, the collaboration and coordination among different cybersecurity teams and departments is required to be aware about the latest insights into global cyber threat risk incident activities and trends, as well as cybercriminals motivations. However, this need the understanding of the multiple facets of the cybersecurity landscape through the following key concerns [29]:

- **What is Happening in CSA:** Detecting whether ongoing IT-risks, Cyber-risks, or Third-party risks exists in organizations digital system's environments or digital infrastructure resources that have been compromised. This also includes the impact of it, which is a part of perception in CSA, and mostly involves automated data gathering tools and pre-processing of the huge amount of gathered digital data. The quantity and quality of gathered digital data by Intrusion Detection System (IDS) and Firewall logs, vulnerability scanning tools, anti-malware log files, and others determine how effective CSA will answer this question.
- **Why is it Happening in CSA:** Monitoring organizations digital system's environment with regard to vulnerabilities, cybersecurity holes, cybersecurity alerts

and others, to become aware of the potential IT-risks, Cyber-risks, and Third-party risks. This also enable understanding the path in which IT-risks, Cyber-risks, Third-party risks evolving, including risk tracking, risk behavior, and needed cybersecurity strategy analysis efficiency, additional reasoning and analysis techniques.
- **What may Happen in Future:** Refers to the ability of forecasting possible future scenarios along with the probabilities, and anticipate damage potential. It includes current situation knowledge and the possibility of its evolution alongside knowledge about the behavior(s) of cybercriminals, which is part of a CSA projection to answer questions such as, what situations may be possible based on organizations current digital system components, cybersecurity posture, and cyber threat risk incidents. This also focus on possible further evolving and exploiting current situations.

Thus, understanding cyber threat risk incidents and the ever-evolving cyber-attack landscape enable organizations to adapt their cybersecurity strategy to better understand their possibly existing vulnerabilities to adapt early appropriate mitigation measures. Hence, a major concern in organizational CSA is the specific knowledge gained, to identify, detect and protect and, if the worst comes to the worst, to defend and recover from cyber threat risk incidents. Nevertheless, the scope of enhancing knowledge in CSA reach different levels of technical aptitude and cybersecurity skills with regard to organizations staff, specialists, managers and decision-makers, responsible for organizational cybersecurity, as well as people who want to take on this responsibility in organizational cybersecurity to defend cyber threat risk incidents. Against this background, results published in [30] showing that between 72% and 95% of cyber threat risk incidents coming from naive cybersecurity practices of staff/employees [31, 32], whereby most cyber threat risk incidents performed by unsupervised use of the Internet, and by means of human errors, where human manipulated to release something valuable, which requires CSA, a cornerstone to ensure that organizations cybersecurity strategy protect in a meaningful way. Given that fact, CSA requires understanding the current organizations cybersecurity situation, and being able to project how the situation will evolve, as an essential part of proactive cyber threat risk incidents defense. However, this also means that CSA includes short-term operational and long-term strategic decision making. As Machine Learning (ML) capabilities are constantly improving in cyber threat risk incident detection and automate response to prevent or mitigate their success to mission-critical services, it also become ad feature in CSA, focusing on innovation and business growth. In this sense, the major parameters of CSA are [33]:

- **Cyber Environment:** Has no borders and is dynamic and highly malleable. Hence, the dynamics and limitlessness of the cyber environment coin a challenge for situational assessment compared to the physical world. Moreover, the spatial properties of the cybersecurity environment mostly are global [34], which makes determination of CSA boundaries complex. Thus, the physical location of organizational cybersecurity of processes, services, and systems is usually used as a spatial boundary for CSA [35],

- **Perception:** Crucial factor in organizations cybersecurity, because it drives trust of economy and citizens to the extent to which one believes that organizational environment is cybersecure for exchanging crucial and critical digital data. However, confidence may be undermined by intruded cyber threat risk incidents to the current organizational cybersecurity situation by means of cyber threat risk incidents defense that evolve as an essential part of cybersecurity to identify, validate, and remediate the exposure as swift as possible. In this regard, perception is responsible for data collection and successful and cybersecure transmission for further processing, which plays a significant role in shaping cybersecurity risk perception.
- **Performance:** Penetration of cyber threat risk incidents in cybersecurity critical environment need adequate measures, which can be relatively small. Thus, the resources required to execute a cyber threat risk incident may scale down at least to an individual cybercriminal actor that only require few prerequisites. Another factor to describe performance is the execution speed of cyber threat risk incidents. The execution speed of cyber threat risk incidents is orders of magnitude faster compared to physical interventions. In this regard, the resources needed for processing of large volumes of digital data are significantly higher in case of CSA [33],
- **Cybercriminal Attackers Advantage:** Cyber threats risk incidents over-take advantages of cyber interventions. The advantages of cyber threat risk incidents include for example anonymity, global reach to probe weaknesses, social engineering to exploit human weaknesses, and the possibility to pick time, place and tools for cyber threat risk incidents [33].

CSA coin an integral part of organizations cybersecurity strategy because it enables to integrate, analyze, evaluate, and remediate the exposure of leaked online credentials as fast as possible, as well as anticipating potential consequences. Nevertheless, cyber threat risk incidents every time find new ways to outwit victims and bypass defenses. However, common cyber threat risk incidents such as phishing, ransomware, malware and others have been around for years, but employees, staff and organizations still don't fully understand or are not aware what they are about. Furthermore, there is even less awareness of newer cyber threat risk incidents such as Quishing, a form of fraud in which cybercriminals attempt to steal sensitive data, similar to phishing attacks. Warning signs in Quishing include unexpected emails describing urgent problems or messages from seemingly trustworthy senders with unusual or surprising requests. Hence, users should be vigilant with QR codes in emails, as these can be an indication of a cybercriminal Quishing attack. Another cyber threat risk incident is Smishing (SMS phishing), which is a cyber threat risk that targets individuals through Short Message Service (SMS) or text messages. Smishing is a combination of SMS and Phishing, while Vishing relates to voice phishing. In case of a dark web intervention, which indicate that cyber threat risk incidents imposed targeting organizations cybersecurity, the cybersecurity awareness it is disappointingly less.

Although cybersecurity skills shortage is not thriving better, organizations requires a proactive priority in identification, analysis and assessment of cyber threat risk incidents, a proactive cybersecurity risk management, and a proactive decision.

1.3.2 Cybersecurity Risk Assessment

Cybersecurity Risk Assessment (CRA) support organizations cybersecurity in understanding the potential cyber threat risk incidents of organizations processes, services, and systems to strengthen operational- and cyber resilience. Thus, CRA is turning into an important narrative of CSA. Given that fact, CRA should be achieved regularly, based on the operational needs, to assess the cybersecurity posture. By conducting the assessments, a baseline of cybersecurity measurements is established. This baseline can be referenced to or compared against future results to further improve the overall cybersecurity posture and resiliency and demonstrate progress. As described in [10, 36] CRA requires several steps being successful. However, CRA is not a meant conducted just once. Instead, the CRA is intended as ongoing determination of organizations cybersecurity measures and should be refined as new technologies like Artificial Intelligence (AI) and new cyber threat risk incident methods become available and adopted. The major steps in Cybersecurity Risk Assessment (CRA) are [10, 37]:

- **Risk Identification and Documentation:**
  - Identify, prioritize and document crucial and critical digital data and assets,
  - Identify, prioritize and document internal and external cyber risks,
  - Identify, prioritize and document internal and external vulnerabilities,
  - Identify, prioritize and document internal processes and records like administrative privileges, activity logs of those granted access, reliance on a managed service provider or a supply chain software vendor's tools, and others,
  - Identify, prioritize and document the access points for each system in use,
  - Identify, prioritize and document misuse of digital data by unauthorized user(s),
  - Identify, prioritize and document digital data recovery process and controls,
  - Identify, prioritize and document the weaknesses in organizations cybersecurity controls,
  - And others.

- **Risk Quantification Analysis and Documentation:**
  - Determine and document the likelihood of cyber threat risk incidents, whereby the likelihood is the probability that a cyber threat risk incident exploited actually, considering the risk incident type, the capability and motivation of the risk incident source, and the effectiveness of controls, as illustrated in Table 1.2.

Table 1.2 Likelihood levels compared to frequency of time per event

Likelihood	Frequency of cyber threat risk incidents
High Likely	Cyber threat risk incidents occur very often, which means *more frequently* than ≥10% of the time per incident
Likely	Cyber threat risk incidents occur often, which means *frequently* in between >1% and <10% of the time per incident
Possibly Likely	Cyber threat risk incidents occur, which means *less frequently* in between >0.1% and <1% of the time per incident
Unlikely	Cyber threat risk incidents occur *rarely*, which means <0,1% of the time per incident

Table 1.3 CIA triad core elements and characteristics

CIA elements	Characteristics
Confidentiality	Preserving authorized restrictions on access and disclosure, including means for protecting personal privacy and proprietary information
Integrity	Maintaining information and data's consistency, accuracy, and trustworthiness over its entire life cycle
Availability	Deploying the fundamental feature of digital systems to prevent data loss successfully

- Determine and document the impact/consequences cyber threat risk incidents have, which compromise the confidentiality, integrity, and availability, as illustrated in the CIA Triad [4], of crucial and critical digital data, as shown in Table 1.3,
- Determine and document what national assets, processes, devices, services and systems are at cyber threat risk incident in the high-risk level, as shown in Table 1.4.
- Determine and document what assumptions qualify for measurement of catastrophic, severe, moderate and low cyber threat risk incident levels, as shown in Table 1.5,
- And others.

- **Risk Evaluation and Documentation:**

  - Prioritize and document the significance of cyber threat risk incident level that compromises,
  - Prioritize and document the cyber threat risk incident level by making use of a m-by-n matrix for determining the risk(s) level(s),
  - Prioritize and document cyber threat risk incident levels potential responses,
  - Prioritize and document what responses are available to counter the different cyber threat risk incident levels,
  - Recommend and document control by using cyber threat risk incident levels as basis to determine action(s) required to mitigate them,
  - Perform and document the personal and/or group that is able to expedite the response time after cyber threat risk incidents,

Table 1.4 Definition of risk levels and potential impact/consequence

Risk Level	Consequences/impact
High Risk	**Not Acceptable Risk Level:** Identified and classified cyber threat risk incident that must be defended in real-time. Potential risk like phishing that try to get access to targeted digital systems, networks and others, intruding malicious codes in targeted objects
Medium Risk	**Moderate Risk Level:** Identified cyber threat risk incidents must be monitored with consideration whether necessary measures done to defend a potential event. Potential risk incidents use malware like ransomware or infect vulnerabilities, and others
Low Risk	**Acceptable Risk Level:** Identified cyber threat risk incidents must be observed to discover changes that could increase the risk level. Potential risk incidents at this risk level are password violations, insider cyber threat risks and others

Table 1.5 Classified event levels with possible impact/consequence

Consequence	Cyber threat risk incident level classification
Unacceptable	Cyber threat risk incident classified *as severe level* risk with successful devastating impact and loss of trust
Severe	Cyber threat risk incident classified as *serious level* risk with serious level of successful impact and loss of trust
Moderate	Cyber threat risk incident classified as *low-level risk* with low level of successful impact that may influence trust
Low	Cyber threat risk incident classified as *no dangerous or marginal risk*

– Perform and document the risk assessment to make appropriate decisions, based on existing measures, current cyber threat risk incident levels, treatment plan(s) and progress status, responses to cyber threat risk incidents, budget.
– And other.

In this regard, the major concern of Cybersecurity Risk Assessment (CRA) is to reduce cyber threat risk incidents and vulnerabilities, protecting organizations digital data and assets, processes, services, systems, and infrastructure resources against unauthorized exploitation. Based on the knowledge gained, strategies on Cybersecurity Risk Management (CRM) be derived.

1.3.3 Cybersecurity Risk-Management

Cybersecurity Risk Management (CRM) is the process that enable identify, assess, prioritize, manage, monitor and respond to cyber threat risks incidents lurking in an organization before they lead to boom. To minimize cyber threat risk incidents, for example, MITRE ATTACK (MITRE ATT&CK) and MITRE DEFEND (MITRE D3FEND) recommends patching and regularly scanning systems for vulnerabilities, two important cybersecurity risk management measures. MITRE ATTACK is a globally-accessible knowledge base of adversary tactics and techniques based on real-world observations of cyber-attacks. Therefore, Cybersecurity Risk Management (CRM) become a vital part of organizations cybersecurity risk management efforts to reduce cyber threat risk incidents to protect organizations assets, data, processes, services, systems, and other resources against unauthorized exploitation, and removes the potential operational, financial, and reputational costs of one too. By not only decreasing cybersecurity risks but increasing overall resiliency, Cybersecurity Risk Management (CRM) proactively enables organizations avoid costly disruptions, bolsters client trust, and safeguards the organizations reputation. Given that fact, Cybersecurity Risk Management (CRM) must be integrated as a periodical process that involves identifying cyber threat risk incidents to organizations data and assets and other critical entities, assessing their likelihood and impact, and implementing measures to mitigate identified cyber threat risk incidents. In this context, NIST SP 800-171 Control 3.11.2 for example requires organizations to scan their digital systems and applications periodically for vulnerabilities, and if new vulnerabilities occur to identify whether those digital systems and applications affected [38]. Also, most cyber insurance providers require organizations to prove they're taking steps to reduce and manage cyber threat risk incidents in order to receive insurance at an affordable rate, or at all [39]. Therefore, the types of most common cyber threat risk incidents organizations should be aware about are for example [39]:

- **Unpatched Software:** Lacks the latest cybersecurity updates, poses significant cyber threat risk incidents, including exploitable vulnerabilities that cyber-attackers can use to gain unauthorized access or cause disruptions.
- **Vulnerable Operating Systems and Browsers:** Significant cybersecurity risks because they can be exploited by cybercriminal actors to gain unauthorized access, execute malicious code, or disrupt operations.
- **Exposed Remote Desktop Protocol (RDP):** Significant cybersecurity risk because it allows remote access to a system over the internet.
- **End-of-life (EOL) Software:** Major cybersecurity risk because it no longer receives updates or support from its vendor.
- **Outdated Protocols and Encryption Technologies:** Pose significant cybersecurity risks because they are vulnerable to known exploits and cyber threat risk incidents.

Based on these requirements the complex CRM process typically requires expertise and involves several essential measures:

- **Identification:** Full visibility across the entire cyber threat risk incident surface is needed to understand any exposures putting an organization at cyber threat risk incidents. Therefore, prioritizing and documenting organizations data and assets that need protection against potential cyber threat risk incidents they may face is required, which includes assessing processes, systems, services, networks, applications, digital data, and others. This measure is crucial because it enable organizations cybersecurity strategy to better understand their potential cyber threat risk incidents surface and prioritize organizations cybersecurity strategic actions and investments.

- **Assessment:** When organizations potential cyber threat risk incidents have been identified, an organization-wide cybersecurity strategy has to assess their likelihood and potential impact, which involves analyzing their vulnerabilities against cyber threat risk incidents that my exploit those vulnerabilities. This measure enables an organization-wide cybersecurity strategy to better understand cyber threat risk incidents potential impact and enable to prioritize mitigation efforts accordingly.
- **Mitigation:** An organization-wide cybersecurity strategy implements the prioritized measures to mitigate identified cyber threat risk incidents based on the assessment done, which include implementing cybersecurity controls, regularly updating hardware and software resources releases, providing regularly cybersecurity staff training of organization-wide cybersecurity strategy, developing incident response and recovery plans and teams, and others, which require a clear understanding of potential cyber threat risk incidents.
- **Monitoring:** CRM is an ongoing process, based on implemented mitigation measures, which requires that organization-wide cybersecurity plan must continuously monitor their processes, systems, services, networks and others for potential cyber threat risk incidents. This enable to ensure that the actions taken for mitigation measures are effective and that new cyber threat risk incidents are identified and addressed promptly. Therefore, monitoring is essential because it enable the organization-wide cybersecurity strategy to maintain their mature cybersecurity posture and respond to new cyber threat risk incidents swiftly. Hence, rapid cyber threat risk incident detection and response keeps organizations covered through the full cybersecurity chain.

Against this background, Cybersecurity Risk Management (CRM) requires monitoring for and prioritizing vulnerabilities, cyber threat risk incidents, and other cybersecurity weaknesses, and then responding to the plethora of alerts generated by basic monitoring or intrusion detection tools. Hence, it offers several benefits to organizations cybersecurity, including [37]:

- **Improved Security:** CRM supports organization's cybersecurity to improve their cybersecurity posture by identifying potential cyber threat risk incidents and implementing measures to mitigate those quickly. Thus, organizations cybersecurity strategy can prevent cyber threat risk incidents and protect organizations digital data and assets through implementing cybersecurity controls.
- **Cost-Effective Security:** CRM is a cost-effective way of managing cyber threat risk incidents to support organizations cybersecurity strategy to prioritize their cybersecurity investments accordingly and allocate their cybersecurity resources effectively. In this regard, prioritizing cybersecurity investments accordingly, organizations cybersecurity achieves the most significant cybersecurity gains with the least resources.
- **Regulatory Compliance:** CRM management support organizations cybersecurity strategy to comply with regulatory requirements such as National Security

Standard [40], Network and Information Security [41, 42], and others, which require organizations cybersecurity strategy to protect their digital data, assets and privacy, to proactively exclude reputational problems.
- **Business Continuity:** CRM support organization cybersecurity strategy to ensure Business Continuity (BC) through identifying potential cyber threat risk incidents and developing contingency plans to address them. An organizations cybersecurity contingency plan primary objective is to protect organizations assets such as processes, systems, services, digital data and others after a cybersecurity breach or disaster has occurred. By developing a cybersecurity contingency plan, which include the essential steps on proactive measures on ways to respond immediately to cyber threat risk incidents, to minimize the impact on organizations cybersecurity strategy operations as well as to prevent future cyber threat risk incidents, breaches, digital data loss and others, to improve organization-wide cybersecurity.

For the improvement of Business Continuity (BC) an actionable Cybersecurity Risk Management (CRM) is required. As indicated in [10, 43], the Recovery Point Objective (RPO) and the Recovery Time Objective (RTO) are two important metrics to evaluate and enhance the maturity level in disaster recovery and cyber threat risk incidents prevention.

The Recovery Point Objective (RPO) primarily refers to the age of data that must be recovered from a backup storage for achieving regular cybersecure operation to resume after a successfully executed cyber threat risk incident, expressed as an amount of time. This time thought of as the point in time before the cyber threat risk incident at which digital data can be successful recovered, which means the time elapsed since the most recent reliable backup. This time can be specified in seconds, minutes, hours or days. Therefore, the purpose of the Recovery Point Objective (RPO) is to determine [10]:

- What is the minimum backup frequency schedule?
- How much digital data can be lost after a disaster?
- How far back the organizations cybersecurity administration team should go to employ sufficient restoration without delaying digital data loss against expected Recovery Time Objective (RTO)?

However, the shorter the time in Recovery Point Objective (RPO) the higher the expected cost for digital data recovery. For example, with a specified Recovery Point Objective (RPO) of 4 h, a new backup at least must be made within 4 h of the last backup. If backups made later, the loss of digital data is too high, and if backups made too frequently, the backup costs increase. Thus, Recovery Point Objective (RPO) determines the loss tolerance, meaning the costs to compensate its potential losses. This specifies how much digital data allowed for loss, and for this loss, the time intervals for digital data backup must be correct [44]. Let's assume a cyber threat risk incident occurs and the most recent digital data backup copy is from 10 h ago. If the standard RPO is 15 h, the backup is still within the bounds, specified by the RPO.

1.3 Cybersecurity

Essentially, Recovery Point Objective (RPO) answers the question *Up to what point in time can the recovery process move tolerably given the volume of data lost during that interval?* [45]. Factors that affect Recovery Point Objective include [46]:

- Maximum tolerable digital data loss for the specific organizational cybersecurity,
- Organizations cybersecurity specific factors dealing with specific sensitive information,
- Organizations digital data security digital data storage options, such as physical files versus cloud storage, which can affect the recovery speed,
- Cost of digital data loss and the loss of organizations cybersecurity operation,
- Compliance schemes that include provisions for digital data recovery, digital data loss, and digital data confidentiality, integrity, and availability, which can affect organizations cybersecurity operations,
- The cost of implementing disaster recovery solutions.

Another factor additionally included in the evaluation of Recovery Point Objective (RPO) is the Time to Data (TtD), which describes the required time to retrieve the backup digital data and deliver it to the restore location [47].

The Recovery Time Objective (RTO) is the maximum acceptable time length that organizations cybersecurity processes, systems, services and others can be down after a successful cyber threat risk incident. Therefore, the Recovery Time Objective (RTO) defines the time span between the occurrence of the cyber threat risk incident and the recovery to regularly operation. Hence, its measures are in seconds, minutes, hours or days [48]. RTO is an essential consideration in a Disaster Recovery Plan (DRP) [49]. A Disaster Recovery Plan (DRP) provide step-by-step instructions to follow during an incident in order to restore business service or business operation. In general, the Disaster Recovery Plan (DRP) outline [50]:

- Nature of the disaster, e.g. cyber threat risk incidents, power outages, hardware failures, and others,
- Information about affected systems, devices, resources and others,
- Any tools or services that can aid in the recovery process, such as backup tools, and others,
- Teams responsible for recovering the failed systems, devices, resources and others,
- Protocols for communicating with team members and documenting recovery steps.

Once an organization has defined the RTO, the cybersecurity team can decide which DRP methods are best suited to the situation, which means that the DRP covers the organizations ability to respond to and recover from an unexpected cyber threat risk incident negative impact. Thus, DRP methods enable organizations to regain use of critical processes, systems, services and IT infrastructure resources as soon as possible after a disaster recovery occurs. Against this background, the DRP process involves [10]:

- Identifying potential cyber threat risk incidents to organizations cybersecurity strategy processes, systems, services and others,
- Establishing procedures for responding to and resolving outages,
- Mitigating the cyber threat risk incidents and impact of outages to operations.

In this context, DRP is a subset of a Business Continuity Plan (BCP), whereby BCP focusses on restoring the essential function of an organization as determined by a Business Impact Analysis (BIA) [51]. A BIA is a useful tool for developing a BCP and an important part of a Disaster Recovery Plan (DRP). By identifying potential failure modes and the associated costs, the BIA provides valuable insights that are then used to formulate a comprehensive DRP. The BIA is conducted after establishing organizations expectations and needs, as well as the organization's business objectives, systematically and reviewed at planned intervals and after significant changes. These play important roles for determining when the impact becomes unacceptable [52].

Calculating RTO requires determining how swiftly the recovery process for a given process, system, service or data needs to happen, after a cyber threat risk incident. Basis for this is the loss tolerance, the respective organization has as part of the BIA, a systematic process to determine and evaluate the potential effect of an interruption to critical cybersecurity operations as result of a disaster, e.g. through cyber threat risk incidents and others. Therefore, a BIA is an essential component of the respective organizations BCP, which includes an exploratory component to reveal any cyber threat risk incidents and vulnerabilities, and a planning component to develop strategies minimizing cyber threat risk exposure, to quantify potential loss of current cybersecurity activity. The result is a BIA report, which describes the potential risks specific to the respective organization [42]. Hence, for cyber threat risk incidents defining the loss tolerance is required, which involves how much operational time the respective organization can afford to lose after a cyber threat risk incident, before regular cybersecurity operations must resume. In this regard, completely inventory of the organizations cybersecurity environment is the first step in the RTO procedure, including all processes, systems, services and other security-critical applications, data and organizational assets. Without an accurate inventory, there is no way accurately determine RTO [49].

Given that fact, the goal for the design of RPO and RTO is to implement a DRP, which in practice means a backup strategy that is as efficient as possible. This backup strategy should bring the disturbed system back to a regular operational or functional state.

In order to design the backup system practically, the possible downtime costs considered together with the running costs for the implementation of the required DRP. It is particularly important to notice that TtD has a negative impact on both, RPO and RTO. The reason is that RPO cannot completed until backup digital data is available for disaster recovery. RTO in disaster recovery can only happen after backup digital data is available. Therefore, Time To Detect (TTD) should be as small as possible when designing RPO and RTO in order to keep the negative impact as small as possible [10].

When designing RPO and RTO to cyber-secure specific organizations processes, systems, services and others, they not necessarily have the same scope. In IT systems, the aim is to implement the smallest possible RPO, as no digital data should be lost. For OT systems, the most important aim is to get the system after disruption back to a functional state as quickly as possible. However, the term functional does not describe achieving 100% of the regular operational state before disruption. An acceptable partial state is when the essential functionality restored and the RTO shortened. The remaining missing features not required for the essential functionality restored afterwards, after the required partial state be achieved [53]. Nevertheless, key recovery objectives that define how long the organization can afford to be offline and how much digital data loss it can tolerate, is determined by [10]:

- **RPO:** Set the maximum amount of digital data losing the respective organization can tolerate, measured in time from the moment a cyber threat risk incident occurs to the last valid digital data backup,
- **RTO:** Set the maximum length of time it takes to restore regular operation following a digital data loss or an outage.

Thus, the DRP defines how, and how fast, to recover from a cyber threat risk incident that unexpectedly renders critical operations and make digital data inaccessible. As such, it prepares for getting back to regular and cybersecure operations quickly, and minimize damage to the organization's cybersecurity.

1.3.4 Cybersecurity Maturity Level Model

Digital data analytics of cybersecurity risk(s) represent a central feature to reach depth knowledge of potential vulnerabilities, cyber threat risk incidents in critical cybersecurity processes, systems, services and others in digital environments, which may need to reassess the cybersecurity level achieved, creating a baseline to ensure the cybersecurity posture is at least maintained, and preferably enhanced. Therefore, evaluating the current status of the situational cybersecurity risk awareness is an essential part in every proactive cybersecurity strategy, to achieve an adequate Cybersecurity Maturity Level (CML). In this regard, the Cybersecurity Maturity Level Model (CMLM) is a structured framework designed to evaluate organizations cyber posture management processes, practices, and controls, and refers to an organization's cybersecurity position relative to its expected cybersecurity risk environment and tolerances, to better protect and defend against potential impacts of cyber threat risk incidents. However, CMLM it's more than just technological aspects [53, 54], equivalent to no uniform model to determine the maturity level, whereby the technological dimension refers to the current state of the situational cybersecurity awareness within the particular cybersecurity processes, systems and others. This include issues associated with critical development paths for organizations cybersecurity, such as: i) having employees that have acquired the necessary know-how and skills, and ii) having the necessary and approved budget [55, 56]. Using the CMLM,

Table 1.6 Maturity levels with assessment criteria

Maturity level	Assessment criteria
0	No activities towards situational cybersecurity awareness and CRA
1	Planning on the subjects of situational cybersecurity awareness and CRA, but no concrete implementation yet
2	Parts of planning situational cybersecurity awareness and CRA have already been implemented
3	Topic situational cybersecurity awareness and CRA is fully implemented and fully documented
4	Topic situational cybersecurity awareness and CRA is continuously checked for the state of the art and efficiency
5	Topic situational cybersecurity awareness and CRA is subject to a continuous improvement process

the underlying dimensions must be operational using evaluation criteria [10] scaling the evaluation criteria for the respective maturity levels. The example shown in Table 1.6 scales the evaluation criteria into six maturity levels.

Table 1.6 scale the assignments of the maturity level evaluation criteria, as illustrated in the so-called radar diagram, also called spider-diagram, in Fig. 1.1.

The 2-D chart in Fig. 1.1 represents multivariate data of quantitative variables mapped onto an axis and plots the data as a polygonal shape over all axes. All axes have the same origin and relative position, which is particularly useful for representing the current maturity state of the objects under investigation. Radar charts plots a series of observations or cases with multivariate data for comparison purposes, whereby a polygon represents each observation or case. However, graphing multiple observations or cases can become messy, because it can be hard to distinguish more than three or more stacked polygons visually.

The radar chart in Fig. 1.1 refers to the NIST Cybersecurity Framework (NIST CSF 1.1), which describes practices for cybersecurity protection, organized in a hierarchy of core functions [57]. The NIST CSF 2.0 core functions are:

- **Govern:** Enables to establish and monitor organizations cybersecurity risk management strategy, expectations, and policy.
- **Identify:** Enables managing cybersecurity risk(s) by asset management in business environments through risk assessment and risk management strategy,
- **Protect:** Enables the ability to limit the impact of potential cyber threat risk incidents through cybersecurity awareness control and training, information protection procedures, data security by protective technology,
- **Detect:** Enables discovery of anomalies and cyber threat risk incidents through continuous cybersecurity risk monitoring and intrusion detection processes,
- **Respond:** Includes appropriate activities like response planning, analysis, mitigation, improvements and communication regarding potential cyber threat risk incidents,

1.3 Cybersecurity

Fig. 1.1 Radar chart is a pentagon with 5 equilateral triangles, called dimensions, that represent e.g. 5 of the six core functions of the NIST Cybersecurity Framework NIST CSF 2.0, called identify, detect, protect, respond, and recover. The 5 equilateral triangle sites of the radar chart are divided into 6 maturity levels, ranging from 0 to 5, that compares the maturity of the different investigated objectives, e.g., employee skills (dotted line), cybersecurity plan (bold line), cybersecurity risk management (dashed line)

- **Recover:** Enables recovery planning back to normal operation, minimize impact of a cyber threat risk incidents by on-site restoration and improvements and emergency communication plan.

Interpreting the representation of maturity levels based on 5 of the 6 NIST CSF 2.0 core function in the radar chart in Fig. 1.1 shows that non-of-them achieve a maturity level of 3 and/or higher. Maturity levels below 3 represent the perseverance of one or more significant obstacles to attaining the objective of maturity level 3. This means for the results, shown in Fig. 1.1, that the responsible cybersecurity personal in the organizations do not have sufficient cybersecurity skills to support 5 of the 6 NIST CSF 2.0 core functions in an adequate way. Therefore, the CMLM refers to the objectives that must be developed further and need action to lift them at least to a mean maturity level of 3 [58].

Another condition, determining the level of cybersecurity maturity, is to estimate the effort that arises from the NIST CSF specific criteria identify and prioritize, assess and analyze, mitigate and control, evaluate and adjust, to achieve a higher maturity level by eliminating identified weaknesses. In this context, the current maturity level is merely a part of an intermediate state to determine the knowledge, skills or activities to acquire the next higher maturity level. Against this background, CMLM also initiate recommendations for further development of the current state in the direction of the strategic *target to-be state* in organizations cybersecurity.

Table 1.7 CMLM criteria for situational cybersecurity awareness culture

Maturity level	Maturity level assignment	Maturity level criteria
1	Basic Compliance	No real situational cybersecurity awareness culture activities; high risk for severe cyber-break likelihood
2	Basic Security Awareness	Basic situational cybersecurity awareness culture; risk for cyber-break likelihood are moderately difficult
3	Programmatic Security Awareness and Behavior	Situational cybersecurity awareness culture programmatic wise implemented; risk for cyber-break likelihood is medium
4	Security Behavior Management	Situational cybersecurity awareness culture is implemented and fully documented; risk for cyber-break likelihood persist at light
5	Sustainable Security Culture	High situational cybersecurity awareness culture; low risk for cyber-break likelihood

Beside the foregoing illustration of the CMLM, applications can also be more specifically generating a maturity model for the specific topic cybersecurity awareness culture in organizations security, as shown in Table 1.7.

From Table 1.7 it can be seen that the situational cybersecurity awareness maturity level 5 depict a low risk. Low cybersecurity risk means there are few anomalies outside the usual concern for cyber-attack initiated cyber threat risk incidents. Hence, at low risk for cyber threat risk incidents likelihood the organizations cybersecurity strategic process, system and service activity are deemed normal, which means no major or dangerous cyber risks occurring. However, for low risk cyber threat risk incidents likelihood specific cybersecurity tools and methods required to avoid a potential successful execution of anomalies outside the usual concern for cyber-attack initiated cyber threat risk incidents. This can be achieved in general by Cybersecurity Risk Management (CRM), Cybersecurity Compliance Management (CCM), Business Continuity Management (BCM), Policy And Documentation Management (PADM), to defend cyber threat risk incidents, vulnerabilities, misuse of passwords, insider or outsider cyber risk, malicious code intrusion and others, to achieve a sustainable and high Cybersecurity Maturity Level (CML) in organization-wide cybersecurity strategic processes, systems, service and others.

1.3.5 Cybersecure by Design and Default

Cybersecure by Design and Default (SDD) is a cybersecurity approach for digital systems and/or product development that fit with customer-centric cybersecurity requirements from the beginning and the delivery, released in a state where customers are secure out-of-the-box. With it, Cybersecure by Design focuses on embedded cybersecurity throughout the entire design process, and Cybersecurity by Default ensures that a digital system and/or product is inherently cybersecure from the start. The Cybersecure by Design principles based on the U.S. Cybersecurity &

1.3 Cybersecurity

Infrastructure Security Agency (CISA) specifications that evaluate organizations objectives, digital electronic systems and cyber risks with regard to understand their vulnerabilities and strengths to make recommendations. With it, Cybersecure by Design prioritize customers cybersecurity as a core business requirement, rather than merely treating it as technical feature to significantly mitigate the number of exploitable flaws, before launching them to the market for widespread use. Hence, it can include cybersecurity features such as Multi-Factor Authentication (MFA), logging, and Single Sign-On (SSO), available at no additional cost.

The Cybersecurity by Design principles are structured by CISA in goals to complement and build on existing software cybersecurity best practices, including those developed by CISA, NIST, other federal agencies, and international and industry best practices. CISA continues to support adoption of complementary measures that advance a cybersecure by design posture [59]. The Cybersecurity by Design pledge by CISA contain seven goals, which e.g. manufacturers are pledging to work towards, in addition to context and example approaches to achieve the goal and demonstrate measurable progress [60], described as follows:

- **Goal 1—Multi-Factor Authentication (MFA):** Within 1 year of signing the pledge, demonstrate actions taken to measurably increase the use of MFA across the manufacturer's products.
- **Goal 2—Default Passwords:** Within 1 year of signing the pledge, demonstrate measurable progress towards reducing default passwords across the manufacturers' products.
- **Goal 3—Reducing entire Classes of Vulnerability:** Within 1 year of signing the pledge, demonstrate actions taken towards enabling a significant measurable reduction in the prevalence of one or more vulnerability classes across the manufacturer's products.
- **Goal 4—Security Patches:** Within 1 year of signing the pledge, demonstrate actions taken to measurably increase the installation of security patches by customers.
- **Goal 5—Vulnerability Disclosure Policy (VDP):** Within 1 year of signing the pledge, publish a VDP that authorizes testing by members of the public on products offered by the manufacturer, commits to not recommending or pursuing legal action against anyone engaging in good faith efforts to follow the VDP, provides a clear channel to report vulnerabilities, and allows for public disclosure of vulnerabilities in line with coordinated vulnerability disclosure best practices and international standards.
- **Goal 6—Common Vulnerabilities and Exposures (CVEs):** Within 1 year of signing the pledge, demonstrate transparency in vulnerability reporting by including accurate Common Weakness Enumeration (CWE) and Common Platform Enumeration (CPE) fields in every CVE record for the manufacturer's products. Additionally, issue CVEs in a timely manner for, at minimum, all critical or high impact vulnerabilities (whether discovered internally or by a Cybersecurity and Infrastructure Security Agency third party) that either require actions by a customer to patch or have evidence of active exploitation.

- **Goal 7—Evidence of Intrusions:** Within 1 year of signing the pledge, demonstrate a measurable increase in the ability for customers to gather evidence of cybersecurity intrusions affecting the manufacturer's products.

As referred to, Cybersecurity by Design principles emphasize cybersecurity throughout software design and development. Cybersecurity by default ensures a zero-day product is inherently cybersecure out-of-the-box. Proper execution of Cybersecurity by Default rests on three fundamental pillars as reported in [60], described as follows:

- **Shift Left:** Focuses on detecting vulnerabilities early in the development process. Developers need to write cybersecure code, avoiding common pitfalls identified in resources like the Open Worldwide Application Security Project (OWASP), which is a nonprofit foundation dedicated to improving software security. For example, the key categories in OWASP Top 10 Vulnerabilities include injection, insecure design, security misconfiguration, vulnerable and outdated components, identification and authentication failures, software and data integrity failures, security logging and monitoring failures, and server-side request forgery. Beside this the OWASP list and highlights the most dangerous software weaknesses that threaten the safe operation of software and are often the root cause of vulnerabilities.
- **Enforcing Cybersecure Configurations:** Ensures consistent cybersecurity across all deployments, regardless of user experience or technical expertise. It also simplifies user experience, since they don't have to make configuration decisions.
- **Cybersecure the Software Supply Chain:** Todays software development procedure has become somewhat as an assembly line that relies heavily on third-party libraries and open-source code. Under cybersecurity by default, developers need to pay strict attention to cybersecurity of these components so they don't introduce vulnerabilities.

Against this background, AI and ML also can show how Cybersecurity by Default principles are incorporated into software development cycles, as illustrated in [61], described as follows:

- **Automated Vulnerability Detection:** AI tools can continuously scan code for vulnerabilities, both known and unknown, so they can be addressed early in a Software Development Life Cycle (SDLC), and/or a Secure Software Development Life Cycle (SSDLC).
- **Proactive Cybersecurity Modeling:** By analyzing attack patterns, AI can predict cyber threat risk incidents; this allows proactive cybersecurity modeling to build software with baked-in defenses against those cyber threat risk incident attacks.
- **Intelligent Developer Assistance:** AI analyze code and make real-time suggestions about secure coding practices to development teams.

1.3.6 Secure Software Development Live Cycle

The Software Development Lifecycle (SDLC) is a structured process that enables high-quality software development. In this regard, Secure SDLC (SSDLC) integrates cybersecurity in the development process, resulting, for example, in cybersecurity requirements being captured alongside functional requirements, cyber risk analysis performed during the design phase, and even cybersecurity testing performed in parallel with development. The SSDLC process contain seven pillars, as described in [62] that are as follows:

- **Requirements:** Defining what the problem is, what the security requirements are, and what the definition of "done" looks like. This is the point where all bug reports, feature requests and vulnerability disclosures transition from a ticket to a project.
- **Planning:** After identifying the problem, determine what the solution is. This need to decide what should be build. As in the requirements phase, the planning phase should involve input and feedback from the organizations cybersecurity team to ensure the solution being proposed solves the problem in a way that is as cybersecure as it is valuable to the customer.
- **Design:** With the security requirements in place, determine how to achieve the designated solution is required, which generally involves designing the solution from end to end. This raise questions such as: What systems will be affected? Which services will be created or modified? How will users interact with this feature? and others.
- **Implementation:** Now the design is implemented into code, and cybersecurity practices come into place. Static analysis is used to run on every commit or push, giving development teams near-real-time feedback about the written code. For the complete code a code review process is necessary.
- **Testing and Deployment:** After code review testing is required, where robust security scanning tools come into play, allowing for an in-depth analysis of cybersecurity of the application. As vulnerabilities are found in this way, solutions can be built into automated tools to protect against regressions in the future.
- **Maintenance:** Generally used to identify and remediate defects in the code, but also for discovering vulnerabilities. From supply chain risks to zero-day exploits, the cybersecurity landscape is ever-changing, and having a process in place to identify and respond to cyber risk problems as they arise is a critical step implementing a cybersecure SDLC.
- **Closed Loop Pillar:** The cybersecure SDLC is a circle, which means that after maintenance it's required to start all over again. Every bug, improvement or vulnerability identified in the testing and maintenance phases will kick off its own requirements phase. Cyber secure software development, as a practice, is a constant cycle of continuous improvement.

Advancing the SDLC needs to reach a next level of a cyber secure SDLC, considering also other specification, such as:

Advancing the SDLC needs to reach a next level of a cyber secure SDLC, considering also other specification, such as:

- **OWASP SAMM:** The OWASP Software Assurance Maturity Model (SAMM) is the successor to the original OWASP Comprehensive, Lightweight Application Security Process (CLASP). It's a robust model that provides clear guidance for integrating cybersecurity practices into the software development process, with an emphasis on tailoring cybersecurity efforts to the appropriate cyber risk profile for an organization.
- **NIST SSDF:** The NIST Secure Software Development Framework (SSDF) is a set of fundamental cybersecure software development practices based on established best practices from cybersecurity-minded organizations. It breaks the SDLC into the following four categories, each aimed at improving an organization's software cybersecurity posture:
  - Prepare the organization: Ensure the people, processes and technologies are prepared to perform cybersecure software development at the organizational level and within individual development teams or projects,
  - Protect the software: Protect all components of the software from tampering and unauthorized access,
  - Produce well-secured software: Produce well-secured software with minimal cybersecurity vulnerabilities in its releases,
  - Respond to vulnerabilities: Identify residual vulnerabilities in software releases and respond appropriately to address those vulnerabilities and prevent similar vulnerabilities from occurring in the future.

1.4 Operational Technology Security

Industry 4.0 (I4.0) is reshaping the production landscape by connected and networked machines, plants and systems applying Operational Technology (OT). OT includes hardware and software used to manage, monitor, and control industrial processes, production equipment, and other activities, which encompasses a broad range of programmable systems and devices that work together with the physical environment or manage systems and devices that interact with the physical environment. In this regard, OT is crucial to many enterprises as it is the backbone of industry productivity. It supports the production activities, services and goods that are delivered to the business' customers. Thus, OT is an especially critical component for asset-intensive organizations, such as critical infrastructure sectors such as energy, manufacturing, transportation, utilities and others. Therefore, OT used in a manifold of application domains such as Industrial Control and Automation Systems (ICAS), physical access control systems, physical environment monitoring and measuring systems, and others. In this context, the information produced through

1.4 Operational Technology Security

OT can also be used to generate insight into the production performance, which allows a more accurate decision-making process regarding productivity, and its business strategy. In turn, this leads to organizations searching for a connection between Information Technology (IT) systems and OT entities, particularly in industries that are seeking real-time analysis of production data to draw conclusions instantly, which allows businesses to react without delays. Unlike IT, which focuses on data processing and communication, OT is concerned with the physical operations of machinery and processes. These systems and devices have brought significant benefits to detect or cause a direct change through monitoring and/or control of devices, processes, increased efficiency and connectivity but it has also introduced new cybersecurity challenges and cyber threat risk incidents.

As described in [63], OT systems consist of combinations of control components, e.g., electrical, mechanical, hydraulic, and pneumatic that act together to achieve a specified objective, e.g., smart manufacturing environments like in Industry 4.0, and others.

Therefore, the control components of OT systems include the specification of the desired output or performance. Hence, OT systems can be configured in one of following ways:

- **Open Loop System:** Output is controlled by established settings,
- **Closed Loop System:** Output affects the input in such a way as to maintain the desired control objective,
- **Manual Mode System:** System is completely controlled by humans.

The several types of common OT systems, including

- **Supervisory Control And Data Acquisition (SCADA):** Used for controlling, monitoring, and analyzing industrial devices and processes. SCADA consists of software and hardware components and enables remote and on-site gathering of data from the industrial components and devices. It enables industrial organizations to remotely manage their industrial applications, because the industrial organization cab accesses the component, device or system digital data, and control them without being on site. SCADA systems are critical as it enable maintain efficiency by collecting and processing real-time digital data. The main components of a SCADA system are Human-Machine Interface (HMI), Remote Terminal Unit (RTU), Programmable Logic Controller (PLC), and SCADA Master Station (SMS). SCADA is a centralized system that monitors and controls the entire area. Typical application areas of SCADA are: Water treatment and distribution, Electrical power grid management, oil and gas production, manufacturing and industrial processes. In this sense, SCADA technology has revolutionized the way industrial organizations operate by providing real-time visibility, control, and automation of complex industrial processes.
- **Distributed Control Systems (DCS):** Computerized system that automates industrial equipment used in continuous and batch processes, while reducing the risk to people and the environment. DCS is integrated as a control architecture that contains a supervisory level of control to oversee multiple integrated subsystems that are responsible for controlling the details of a localized process. It

uses a centralized supervisory control loop to mediate a group of localized controllers that share the overall tasks of carrying out an entire production process.
- **Programmable Logic Controller (PLC):** Used in SCADA and DCS systems as control component of an overall hierarchical system to locally manage processes through feedback control. PLC can also be implemented as primary controller in smaller OT system configurations to provide operational control of discrete processes such as assembly lines, process controllers, and others.
- **Physical Access Control System (PACS):** Type of physical security system designed to control access to an area. Unlike standard physical barriers, physical access control can control who is granted access, when the access is granted, and how long the access should last.
- **Industrial Internet of Things (IIoT):** Consists of sensors, instruments, machines, and other electronic devices that are networked together and use internet connectivity to enhance industrial and manufacturing business processes and applications. IIoT also includes ML and big data technology along with PLC, SCADA, Remote Terminal Units (RTU), Human-Machine Interface (HMI), Intelligent Electronic Devices (IED), and automation technology by means of implementing self-diagnosis and rectification capability. As IT and OT systems continue to converge and become even more interconnected, control of physical processes remains relatively related to OT. However, IIoT is the next consequent step of innovation influencing the way connecting and optimizing systems, and others.
- And Others.

OT systems are vital for the operation of critical infrastructures, which are often highly interconnected and mutually dependent systems, both physically and through a host of information and communications technologies [63]. Given that fact, critical infrastructures are often referred to as *system of systems* because of the interdependencies that exist between various industrial sectors and the hyper connections between business partners [64, 65]. Therefore, cyber threat risk incidents in one infrastructure can directly and indirectly affect other infrastructures through cascading and escalating malicious failures, which dramatically highly increase the demands on OT cybersecurity in order to effectively counter cyber threat risk incidents. Thus, cybersecurity remains a top priority for all organizations, as high-profile hacking cyber threat risk incidents have highlighted the cyber threat risk incidents events in today's highly networked industrial environments and infrastructure resources. However, cybersecurity is more than installing an antivirus program. Cybersecurity is the application of controls, practices, policies, processes, services, strategies and technologies, designed to protect digital systems, critical and crucial infrastructure, networks and information systems and other from a wide range of cyber threat risk incidents.

Cybersecurity itself is a computer science-based discipline, which deal with the presence of adversaries and their cyber threat risk incidents repertoire through the Internet. Within computer science, cybersecurity spans many areas, including (but not limited to) data security, cryptography, hard- and software security, network and systems security, as well as cybercrime, cyber-forensic research and other topics.

Thus, cybersecurity is fundamental to both, protecting and defending digital systems, micro-electronic devices and systems like OT systems, mobile devices, networks, servers and other from malicious cyber threat risk incidents. In this context, cybersecurity breaches through cyber threat risk incidents become reality and all organizations face a constant danger of (malicious) cyber threat risk incidents, because routinely try to convince authorized users to hand over their credentials, e.g., through phishing mails and legitimate-sounding phone calls among other sophisticated tactics like AI based deep fake, a crucial and critical security problem of today which require to develop new and enhanced solutions for cybersecurity. Therefore, end users in all organizations must be aware of suspicious email attachments, rouge websites, unauthorized applications, and others, which helps to avoid that cyber threat risk incidents be executed successfully and damage or manipulate internal processes through an unauthorized access. Additional examples of the potential impact of an OT cyber threat risk incident are listed below:

- Impact on national security,
- Reduction or loss of production capability at one site or multiple sites simultaneously,
- Damage to equipment,
- Release, diversion, or theft of hazardous materials,
- Environmental damage,
- Violation of regulatory requirements,
- Product contamination,
- Criminal or civil legal liabilities,
- Loss of proprietary or confidential information,
- Loss of brand image or customer confidence,
- And others.

Thus, cybersecurity breaches mostly have negative impacts that last longer than other types of incidents on all stakeholders, including employees, shareholders, customers, and communities in which an organization operates. Hence, senior management should identify and evaluate the highest priority items to estimate the annual business impact, e.g., in financial terms, reputation and others.

Nevertheless, one of the most problematic elements of cybersecurity is the fast and constantly evolving nature of cybersecurity breaches increased over time, which requires the fast and constantly evolving nature and expansion of cybersecurity defense methods. The traditional approach in IT security focus on resources on the most critical data, assets and processes to protect them against the biggest known cyber threat risk incident, which leaves some less important assets disregarded.

Such an approach is insufficient for the current advancement in digitalization and new technologies in digital transformation organization face for their OT systems, networks and infrastructure resources. Given that fact, cybersecurity professionals assume that the traditional approaches to OT cybersecurity can become unmanageable because the progress in the cyber threat risk incident landscape has become impossibly complex and unprecedented. However, manual and semi-automated cybersecurity monitoring checks and interventions can't keep up with the constantly evolving cyber threat risk incident landscape. In this regard, OT cybersecurity identified as vulnerable to cybercrime due to the omnipresent accessibility and connectivity, which makes them vulnerable cyber threat risk incidents.

Given that fact, cybersecurity teams worldwide are trying to analyze vulnerabilities in order to gain deeper knowledge about them, to build up upon efficient and effective cybersecurity strategies, like the National Institute of Standards and Technology Cybersecurity Framework (NIST CSF) with its five core functions: Identity (ID), Protect (PR), Detect (DE), Respond (RE) and Recover (RC). Their application for OT systems is reported in a NIST special paper [63].

Comparing IT systems with OT systems many characteristics found that differ from each other, including different risks, priorities, performance, reliability requirements and application, which may be considered unconventional in a typical IT environment. In this context, cybersecurity protections must be implemented in a way that maintains OT system integrity during normal operations as well as during cyber threat risk incidents. Some special considerations in OT systems cybersecurity include [63]:

- **Timeliness and Performance Requirements:** OT systems are generally time-critical with the criterion for acceptable levels of delay and jitter dictated by the individual installation. Many OT systems utilize Real-Time Operating Systems (RTOS), where real-time refers to timeliness requirements.
- **Availability Requirements:** Many OT processes are continuous in nature which means that unexpected outages of systems that control industrial processes are unacceptable, because OT systems often cannot be stopped and started without affecting production.
- **Risk Management Requirements in OT Systems:** Include safety, fault tolerance to prevent the loss of life or endangerment of public health or confidence, regulatory compliance, loss of equipment, loss of intellectual property, or lost or damaged products.
- **Physical Effects:** OT systems are field devices, e.g., PLCs, operator stations, DCS controllers, which are directly responsible for controlling physical processes.
- **System Operation:** Require different skill sets, experience, and levels of expertise in OT.
- **Resource Constraints:** OT systems with their RTOS and their real-time operating systems (RTOSs) are often resource-constrained systems that do not include typical contemporary IT cybersecurity capabilities.
- **Communications:** Communication protocols and media used by OT environments for field device control and intra-processor communication are typically different from IT environments and may be proprietary.
- **Change Management:** Paramount to maintaining the integrity of both IT and OT systems. Unpatched software represents one of the greatest vulnerabilities to a system. Software updates on IT systems, including cybersecurity patches, are typically applied in a timely fashion based on appropriate cybersecurity policies and procedures.
- **Managed Support:** Different because OT systems service support is sometimes only available from a single vendor.

1.4 Operational Technology Security

- **Component Lifetime for OT Systems**: Often in the order to 10–15 years because technology has often been developed in many cases for specific uses and implementations.
- **Component Location:** Most IT components and some OT components are located in business and commercial facilities that are physically accessible by local transportation.

Some of the typical differences between IT and OT systems are summarized in Table 1.8 [63].

Table 1.8 Summary of typical differences between IT and OT systems

Category	IT	OT
Performance Requirements	*Non-real time.* Response must consistent. High throughput demanded. High delay and jitter acceptable. Emergency interaction less critical. Tightly restric-ted access control be implemen-ted to degree necessary for cybersecurity	*Real-time.* Response time-critical. Modest throughput acceptable. High delay and/or jitter unacceptable. Response to human and emergency interaction critical. OT access should strictly control but should not hamper or interfere with human-machine interaction
Availability (Reliability) Requirements	Availability deficiencies can often be tolerated, depending on the system's operational requirements	Availability requirements may necessitate redundant systems Outages be planned and scheduled days or weeks in advance. High availability requires exhaustive pre-deployment test
Risk Management Requirements	*Manage data.* Data confidentiality and integrity is paramount. Fault tolerance is less important Major risk impact is a delay of business operations	*Control physical world.* Human safety paramount, followed by process protection. Fault tolerance essential; even momentary downtime unacceptable. Major risk impacts regulatory non-compliance, environmental impacts, loss of life, equipment, and production
System Operation	Systems designed for use typical OSs. Upgrades straightforward with the availability of automated deployment tools	Systems often use different and possibly proprietary OSs, sometimes without security capabilities built in. Software changes be carefully made, usually by software vendors, because of specialized control algorithms, potentially modified hardware and software involved
Resource Constraints	Systems specified with enough resources to support addition of third-party applications, such as cybersecurity solutions	Systems designed support intended industrial process, may not have enough memory, computing resources to support addition of security capabilities
Communications	Standard IT communications protocols used. Primarily wired networks with localized wireless capabilities. Typical IT networking practices employed	Many proprietary and standard communication protocols used. Several communications media used, including dedicated wired and wireless. Complex network exist that sometimes require the expertise of control engineers
And others	…	…

Operational Technology Security (OTSEC) enables cybersecurity teams to fine-tune traditional, digitalized and new technology-based OT processes, reducing the potential of cyber threat risk incidents and safeguarding them against malware-based cyber risks. This requires an effective and efficient OT cybersecurity strategy to prevent the inadvertent or unintended exposure of critical and sensitive data and infrastructure resources, to enable industrial organizations to prevent their OT system-based process and production activities, capabilities and future intensions from cyber risks. Leaking such information, cyber threat risk incidents be able to cause major damage like commit identity fraud or theft if employees reuse their login credential across multiple online services [63, 66]. A best practice in OTSEC analysis contains a five-step approach, as shown in Table 1.9.

Digital transformation require that OT systems work together with IT systems. OT network components and industrial networks are connected to IT network

Table 1.9 OTSEC actions with capabilities

OTSEC task	Capability of OTSEC task
Identify Sensitive Assets	Identification, classification and periodization of assets of OT systems and devices. This include detailed shop floor information, customer information, intellectual properties of production and product research information. Vital for organizations to focus their resources on protecting this critical data
Identify Possible Cyber Threat Risk Incidents	Analysis of data traffic to determine potential cyber threat risk incidents to OT systems. This include continuous behavioral analysis in OT networks to provide cybersecurity teams with actual and accurate data about what is happening in the network in the context of *what, where, when, who, how*. Data from known and unknown cyber threat risk incidents must continuously collected for this analysis
Analyze Vulnerabilities	Analyze potential vulnerabilities in OT systems to prevent cyber threat risk incidents materialize. This requires a successful asset inventory management solution, based on high quality content of assets, to achieve a strong vulnerability analysis, prioritization, timely remediation of vulnerabilities, and finally tracking the vulnerability management process. This also involves assessing the processes and technological solutions that safeguard data and identifying loopholes or weaknesses that cybercriminals could potentially exploit
Analyze Threat Event Attack Level	Each identified vulnerability has a cyber threat risk incident level attributed to it. The vulnerabilities ranked on the likelihood of cyber-criminals targeting them, the level of damage caused if exploited and the amount of time and work required mitigating and/or repair the damage. The more damage is inflicted and the higher the chance of a cyber threat risk incident to occur, the more resources organizations should place in mitigation cyber threat risk incidents
Plan to Mitigate Cyber Threat Risk Incidents	Provides organizations with everything needed to plan to mitigate the cyber threat risk incidents identified. The final step in OTSEC is putting measures in place, to eliminate cyber threat risk incidents, and mitigate cyber threat risk incidents. Typically, this include updating e.g. operational hardware, creating policies around safeguarding sensitive data and providing employee training on cybersecurity best practices and corporate data policies

1.4 Operational Technology Security

components such as processors, data storage and systems management. With an OT-IT integration, the digital data collected from physical and IIoT devices can also be used to detect cybersecurity effects or increase systems efficiency. IIoT devices are digital systems that use smart sensors, actuators and other devices, such as Radio Identification tags (RFID), to enhance manufacturing and industrial process capabilities. IIoT are networked together to provide digital data collection, exchange and analysis. Insights received from this process aid in more efficiency and reliability in digital transformation environments like I4.0.

Digital systems perfectly integrate computation with real processes, and provide abstractions, modeling, analysis, and design techniques for their overall advanced digital technology-based conceptualization. Their integrated computational and physical capabilities interact Over-the-Air (OTA) by wireless devices, connecting the cyber with the real physical systems, and processes through new modalities in digital transformation dedicated tasks. However, the consolidation of cyber and real physical components within the digital transformation enable new categories of vulnerabilities, with regard to interception, replacement, or removal of data from the communication channels, which result in malicious attempts by cyber threat risk incidents to capture, disrupt, defect, or fail the OT system operations and others. This requires tracing the reason for this new vulnerability to understand the way in which the cyber and the real physical components of OT systems are integrated. In this vulnerable space, the cyber-component provides computing capability, processing, control software and sensory support to facilitate the analysis of data received from various sources of the organizations digital system resources. However, accessibility provides an entrance for launching cyber threat risk incidents that can result in:

- Defective operation if the cyber threat risk incident affects the OT systems,
- Denial of Service (DoS), which is common in the cyber-domain,
- Digital data destruction is the process of destroying digital data so that it's unreadable and can't be accessed or used,
- Digital data exfiltration means unauthorized transfer of digital data from a digital system,
- Digital data corruption means that errors in computer data introduce unintended changes to the original digital data
- And others.

Therefore, cyber threat risk incidents not only have tremendous impact on the probably attacked entity, but can also affect all cyber and physical system entities. In this context, the defense against cyber threat risk incidents is an essential must. In this context, cybersecurity is one of the crosscutting issues in digital transformation. Therefore, in OT cybersecurity it is fundamental that authorized messages delivered at any time, at any place, at the right time in real-time, without any disturbance and without malicious incidents. However, a well-defined use case for an OT cybersecurity plan is essential to ensure the long-term commitment of the organization and the allocation of resources needed for the development, implementation, and maintenance of the cybersecurity program. Hence, the first step in developing an OT cybersecurity plan is to identify the organization's business objectives and missions, as well as how the cybersecurity plan can lower risk and protect the organization's ability to perform those objectives and missions. The business case should capture the business concerns of senior management and provide the business impact and financial justification for creating an integrated organization-wide cybersecurity program. It should include detailed information about the following issues [63]:

- Benefits of creating an integrated cybersecurity plan,
- Potential costs and failure scenarios if an OT cybersecurity plan is not implemented,
- High-level overview of the process required to implement, operate, monitor, review, maintain, and improve the information security program.

The costs and resources required to develop, implement, and maintain the cybersecurity plan should be considered. The economic benefits of the cybersecurity plan may be evaluated in the same way that worker health and safety plans are. However, a cyber threat risk incident on the OT systems could have significant consequences that far exceed monetary costs [63].

Given that fact, it is important to note that the ability to hack is a skill, involving the manipulation of technological assets in some specific way or form. Those, who can successfully hack, called hacker, suggesting that they have the capability to hack. To differentiate between benign and malicious hackers, the hacker community introduced the term cracker, to characterize people who engage in cybercriminal or unethical acts using hacking technologies (see Sect. 1.5). This term suggests that a cracker is different from a hacker and has be treated accordingly. Thus, information security or more in general cybersecurity are essential needs that focus on measures to protect authorized messages from malicious cracker incidents.

The fundamental objective in information cybersecurity refers to protecting information and information systems from unauthorized access, destruction, disclosure, disruption, modification or usage. Therefore, the three fundamental principles in information cybersecurity are Confidentiality, Integrity and Availability, commonly referred to as CIA Triad. This is a model to guide policies for cybersecurity in organizations, which also form the main objectives of any basic cybersecurity program. Sometimes, the model also refers to as AIC Triad, which means availability, integrity and confidentiality, to avoid confusion with the CIA term that also refer to the U.S. Central Intelligence Agency.

1.5 CIA Triad

The CIA Triad is a widely accepted model in information cybersecurity. Information cybersecurity is a compliance that requires industrial, public and private organizations to use policies, processes and procedures protecting and acting against cyber threat risk incidents. In this context, the CIA triad is the cornerstone of information cybersecurity used to represent the three core principles of cybersecurity:

1.5 CIA Triad

Confidentiality, Integrity, and Availability. These principles serve as the bedrock upon that construct defenses against cyber threat risk incidents, ensuring that data remains both cybersecure and accessible when needed. Thus, the CIS triad is a common model that build the basis for the development of cybersecurity systems, used to find vulnerabilities and methods for creating solutions. Against this background, the CIA triad guide the cybersecurity procedures and policies in industrial, public and private organizations, as illustrated in Table 1.10.

The holistic model approach to cybersecurity of the CIA triad address not only the protection of data but also its integrity and availability, which ensures that organizations can maintain both privacy and operational efficiency, along with required solutions in the area of digital systems, data, infrastructure, network and information security, and others. It contains policies and procedures that are basis for a successful cybersecurity protection solution, and adapts to changing cyber threat risk incident possibilities, to withstand and recover rapidly from cybercriminal activities. However, an effective cybersecurity program requires a strategic approach, because it provides a holistic plan how to achieve and sustain a desired level of cybersecurity maturity against internal and external cyber threat risk incident scenarios against the CIA triad. In Table 1.11 cyber threat risk intentions classified in the context of internal and external activities, with regard to specific skills gained, working in industrial organizations, as well as contractors or business associates.

Insider cyber threat risk incidents originate from individuals within an organization who have authorized access to data and information systems. Insiders can be employees, contractors, or partners. Insider cyber threat risk incidents are categorized into intentional and unintentional. Intentional cyber threat risk incidents are driven by various motivations like financial gain, personal grievances, ideological beliefs and others. For example, an employee might exfiltrate proprietary data to a competitor or disrupt information systems due to a revenge attitude or other personal settings. In this context, common methods employed by malicious insiders include data exfiltration through unauthorized email accounts, deployment of malware, alteration of critical data to harm organizational integrity and others like bypassing security controls, using privileged account passwords to exploit vulnerabilities, or employing social engineering tactics to gain further unauthorized access. Defending insider threat risk incidents are based on Machine Learning (ML) techniques to accurately evaluate insider cyber threat risk incidents across organizations credentials, IPs, and devices. The current generation of Security Information and Event Management (SIEM) systems with ML enhancements, offering data enrichment with fingerprinting and Threat Intelligence (CI) context, and new analytical possibilities for detecting insider cyber threat risk incidents. To enable this, a SIEM with ML enhancement include the following capabilities: behavioral baselining, peer group analysis, privileged account analysis, shared account analysis, and locked account analysis. Another option in cyber-securing organizations accounts can be achieved through strategies like Privileged Access Management (PAM), which is crucial to mitigate potential cyber threat risk incidents. Against this background, insider risk management enables to detect, investigate, and address internal cyber risks, such as IP theft, fraud, data leakage and others. Unintentional insider

Table 1.10 CIA triad core elements and capabilities

Core elements	Characteristics of core elements
Confidentiality	Vital cybersecurity characteristic in the digital transformation era, which means protection of sensitive data from unauthorized access and data misuse, ensuring that only those with the necessary permissions can access data and information, e.g., by a set of rules that limits access. Measures undertaken to ensure confidentiality to prevent sensitive data from reaching wrong users, making sure that only right users get it. In a world where sensitive data and information can be weaponized for nefarious purposes, maintaining confidentiality of such data and information is still paramount. Hence, different kinds of methods used to preserve confidentiality, including access control mechanisms and protection as ongoing monitoring, data encryption and secure authentication protocols, testing and training, as common methods of ensuring confidentiality. In this context, user IDs and passwords constitute a standard procedure. Other options include biometric verification, as well as security token, which is a small hardware device that the owner carries to authorize access to a network service, and key fobs, which are small, programmable hardware device that provides access to a physical object, or soft token, a software-based security token, that generates a single-use login PIN. However, to satisfy the desired cybersecurity requirements the solution should include a holistic consideration
Integrity	Fundamentally the assurance that data remains accurate, complete, consistent, and trustworthiness in storage and during transmission. This principle aims to prevent unauthorized alterations or corruption of data, whether intentional or accidental. To maintain data integrity, organizations may implement digital signatures, hashing algorithms, and data backups to verify data authenticity and facilitate data restoration when necessary over its entire life-cycle. This covers the topics data integrity and system integrity. Data integrity is the requirement that information, data, and programs changed only in a specified and authorized manner, while system integrity refers to the requirement that a system performs its intended function in an unimpaired manner, free from deliberate or inadvertent unauthorized manipulation. Against this background, a deficiency in integrity can allow for modification of information, data, and programs stored on the digital systems memory, which can affect the crucial and critical operational functions of digital transformation applications, without ad hoc detection
Availability	Ensures that data and systems are accessible to authorized users whenever needed. This principle is crucial in maintaining the continuous operation of vital systems and services of organizations, even in the face of cyber threat risk incidents. To guarantee availability, organizations may adopt measures such as redundant systems, load balancing, and robust disaster recovery plans. This ensure rigorous maintaining all system hardware, immediately performing hardware repairs when needed, and maintaining correct functioning operating system environment, free of software conflicts. If crucial and critical operational systems cannot access the needed data when required, the information, data, and programs of operational systems are not cybersecure. Thus, availability is a fundamental feature of a successful deployment of digital systems in the era of digital transformation. To prevent data loss, a backup copy may be stored in a geographically isolated location, perhaps even in a digital safeguard. Extra cybersecurity equipment or software such as firewalls or proxy servers can guard against downtime and unreachable data, information, and programs in case of malicious activities such as Denial-of-Service (DoS) attacks, and network intrusions

1.5 CIA Triad

Table 1.11 CIA Triad and respective cyber-attacker's skill background

Confidentiality		Integrity		Availability
Internal	External	Internal	External	Possibilities
Hacker: *Insider:* Accessing internal information as unauthorized user. Downloading/Exporting internal information.	**Cracker:** *Malicious:* Stealing internal secret information	**Hacker:** *Insider:* Maliciously creating, deleting, modifying secret internal information	**Cracker:** *Malicious:* Creating, deleting, modifying secret internal information	**Cracker/Hacker:** *Ransomware:* Attack rendering data unusable until backups are accessed or encryption key obtained
Employee: *Insider:* Unencrypted thumb drive with internal secret organization information			**Hacker:** *Vendor:* Modifying secret internal information	**Cracker:** *Malicious (DoS):* Degrading network performance & affecting organizations operations
				Cracker/Hacker: *Server Failure:* At organizations or vendor site
				Cracker: *Remote Access:* Sneak remotely in attacked network setting up phishing scams, duping users downloading malware-files, executed to commence cybersecurity threat like ransomware,...

cyber risk typically stems from human errors or negligence. Given that fact, sensitive data might be exposed due to improper handling or failure to apply essential security patches. Misconfiguration of network devices can also lead to vulnerabilities and cybersecurity breaches, emphasizing the need for robust PAM. Understanding these cyber risk threat types and their underlying causes is crucial for developing targeted mitigation strategies. Intentional and unintentional cyber threat risk incidents can both result in severe consequences, including data breaches, financial losses, damage to the organization's reputation and others. Therefore, organizations must set out to mature their cybersecurity strategy, because managing third-party cyber risks can become overwhelming, especially as they need to incorporate more cloud-based vendors, to support streamline business operations. This requires to pivot organizations cybersecurity programs to be proactive, with regard to third-party cyber risk management policies and procedures, to act foundational as guidelines for creating a cybersecurity risk management strategy for vendors, which

Table 1.12 Third-party cybersecurity risks, characteristics and capabilities

Third party cybersecurity risk	Characteristics and capabilities
Compliance Risk	Industry standards and regulations often incorporate third-party vendor risk as a compliance requirement, ensuring that the own organizations apply the organizations risk tolerance to their third-party business partners. If a primary control within the own organization is to update security patches every 30 days, then they should hold third-parties accountable to that standard and monitor, to verify their controls´ effectiveness
Operational Risk	Potential risks by third parties can achieve through a technology integrated to provide continued business operations. If the third party experiences a cybersecurity threat and/or cyber-attack that shuts down the service, the own organization may experience business interruption
Reputational Risk	While operational risk applies to the own organizations businesses ability to provide customers a service or product, reputational risk applies to the fact how customers view their partner organization. If the own organization experience a third-party data breach, then the own organization may experience decreased customer trust or loyalty in the aftermath (reputation loss)

reduces potential negative impact on organizations critical networks and information systems and infrastructure resources. The reason for this strategy is that third parties may pose a variety of cybersecurity risks [67], as shown in Table 1.12.

Ideally, when all three CIA triad pillars have been met, the cybersecurity profile of an organization is stronger and better prepared to handle cyber threat risk incidents.

1.5.1 Linking CIA Triad Pillars to NIST Incident Response Lifecycle

The CIA triad represent the three pillars of information security:

- Confidentiality: Preserving authorized restrictions on information access and disclosure, including protecting personal privacy and proprietary information,
- Integrity: Guarding against improper information modification or destruction and ensuring information non-repudiation and authenticity,
- Availability: Ensuring timely and reliable access to and use of information.

In [68] the connection between the three CIA triad pillars and the NIST Incident Response Lifecycle, which is crucial in developing a comprehensive and effective cybersecurity strategy, is described. The description creates a robust cybersecurity posture, with the CIA triad pillars guiding the objectives of the NIST Incident Response Lifecycle.

The *CIA* triad pillar *Confidentiality* is deeply intertwined with the *NIST Incident Response Lifecycle*, as it is a key objective during each stage of the process. The

1.5 CIA Triad

NIST Incident Response Lifecycle breaks incident response into the phase's preparation, detection and analysis, containment, eradication and recovery as well as post event activity. In the *preparation* phase, organizations must establish policies and procedures that prioritize the protection of sensitive data and information, which also is a to do in *NIS2* Article 21.2.a. During *detection and analysis*, identifying breaches of confidentiality is crucial, as unauthorized access to sensitive data and information can have devastating consequences, which also is a to do in *NIS2* Article 21.2.b. Furthermore, during the *containment, eradication*, and *recovery* phase, efforts must be made to prevent further exposure of sensitive data and information and to restore the confidentiality of compromised data that also is a requirement in *NIS2* Article 21.2.

The *CIA* pillar *Integrity* is also closely linked to the *NIST Incident Response Lifecycle*, as maintaining data accuracy and consistency is essential throughout the entire process. During the *detection and analysis* phase, the integrity of log files, system data, and evidence must be preserved to ensure an accurate assessment of the situation, which also is a requirement in *NIS2*. In the *containment, eradication, and recovery* phase, steps must be taken to restore the integrity of any corrupted data and to prevent further unauthorized alterations, which also is a to do in *NIS2* Article 21.2.c and e. Finally, in the *post-incident activity* phase, lessons learned should focus on strengthening the organization's ability to maintain data integrity in the face of future cyber threats risk incidents, which also is a to do in *NIS2* Article 21.2.b.

The *CIA* pillar *Availability* is a critical objective of the *NIST Incident Response Lifecycle*, as it ensures that vital systems and services remain accessible to authorized users even in the midst of a cybersecurity incident. In the *preparation* phase, organizations must develop robust disaster recovery plans and implement redundant systems to guarantee availability, which also is a requirement in *NIS2* Article 21.2.c. During *detection and analysis*, identifying cyber threat risk incidents to system availability is a key priority, as downtime can have severe operational and financial implications, which also is a requirement in NIS2 Article 21.2.b. In the *containment, eradication, and recovery* phase, restoring affected systems to their normal operational state is a primary goal, to minimize downtime and maintaining system availability that also is a requirement in *NIS2* Article 21.2.b, c, and e.

In [68] also *examples of the NIST Incident Response Lifecycle addressing CIA Triad breaches given*, where the NIST Incident Response Lifecycle has been effectively employed to address breaches of the CIA Triad principles. One example is the response to the WannaCry ransomware attack, which primarily targeted the Availability principle by encrypting files and demanding a ransom for their release. By employing the NIST Incident Response Lifecycle, organizations were able to swiftly detect, contain, and recover from the attack, restoring system availability and mitigating the potential damage. Another example is the response to the SolarWinds supply chain attack, which compromised the confidentiality and integrity principles by infiltrating the networks of numerous organizations. Through the diligent application of the NIST Incident Response Lifecycle, affected organizations were able to identify the breach, assess its scope, and take steps to remediate the issue, ultimately restoring both the confidentiality and integrity of their data [64].

To ensure the protection of confidentiality, integrity, and availability, NIST specifies security requirements in multiple security-related areas, which represent a broad-based, balanced information security program that addresses the management, operational, and technical aspects of protecting federal information and systems.

1.6 Cybersecurity Is Paramount

Today information is mainly kept in digital databases, because storing data digitally is both efficient and cost-effective. However, organizations must ensure to use efficient and effective cybersecurity strategies, frameworks and intelligent methods that they not get infected by cyber threat risk incidents, which could dramatically damage organizations workflows and reputation. Therefore, organizations need to further invest in their cybersecurity infrastructure, and execute an educational cybersecurity awareness campaign, to say the least. However, the requirements for cybersecurity differ in a number of ways to enable safe and reliable operations. Within today's, expanded protection sphere an enhanced cybersecurity maturity level is needed (see Sect. 1.3.4), focusing on the central objectives of the CIA Triad (see Sect. 1.5). Hence, tools are necessary to monitor digital systems, network and information systems, infrastructures resources and others, supporting detecting cyber threat risk incidents to respond proactively to them. In this context, threat intelligence, an evidence based actionable knowledge approach, enable supporting cybersecurity. Furthermore, the method of intrusion detection and prevention of cyber threat risk incidents requires algorithms that identifies and analyze suspicious or malicious data, generated by the various cyber threat risk incidents on critical computer systems, network and information systems, infrastructure resources, and others, to mitigate and/or defend them.

Once a sophisticated suspicious or anomalous activity has been identified, a cybersecurity risk assessment (see Sect. 1.3.2) must be carried out to determine the likelihood of impact, and the possible consequence of this cyber risk. This approach relates to risk quantification that evaluates the identified cyber risk to respond actively to the identified cyber threat risk incident in real-time to mitigate or defend is. In this regard, cybersecurity is the probability that a cyber threat risk incident provides serious harm under specific conditions. Given that fact, a cyber threat risk incident is a combination of two factors:

- Probability that a harmful cyber threat risk incident occurs,
- Consequences of the harmful cyber threat risk incident.

Various statistical techniques used to evaluate and quantify cyber threat risk incidents. Furthermore, the objective of quantification is to establish a way ranking cyber threat risk incidents in the order of likelihood and consequences. In the context of likelihood and probable consequence of cyber risk levels qualitative cyber risk analysis is applied. Qualitative cybersecurity risk analysis is appropriately early

1.6 Cybersecurity Is Paramount

activated in a cybersecurity strategy project and is effective in categorizing which potential cybersecurity risks should be combatted or not, and what action must immediately be applied. Qualitative analysis techniques based on classification of acceptance criteria of acceptable risk levels, in conjunction with cybersecurity criteria referred to in the CIA Triad (see Sect. 1.5). An acceptable cyber risk refers to the likelihood of a cyber threat risk incident, whose occurrence probability is low and whose consequences are so slight, that the responsible cybersecurity manager decide to monitor the cyber threat risk incident behavior before combatting it. However, the acceptance criteria have to assure the cybersecurity requirements for accessing a crucial digital device. Some acceptance criteria examples in relation to the CIA Triad shown in Table 1.13.

As shown in Table 1.13, the qualitative cyber risk analysis does not give precise values about the cyber risk, but is effective when only little time is available to evaluate potential cyber risks before they actually happen. However, understanding the impact and potential consequences of cyber risks requires a solid understanding of cyber risks and vulnerability types to decide about likelihood and impact. Nevertheless, many cyber risk models suffer from vague, non-qualified outputs based on partial information or on unfounded assumptions. Against this background, threat intelligence provides an approach that enable cyber risk models to define cyber risk measurements and be more transparent about their assumptions, variables and outcomes, helping answering questions such as:

- Which cyber threat risk incident actors using this specific incident and do they target the business or industrial operational production processes?
- Is the cyber threat risk incident observed found more often?
- Is the trend of this cyber threat risk incident up or down?
- Which vulnerabilities does this cyber threat risk incident exploit and are those vulnerabilities present in business or industrial operational production systems?

Table 1.13 Security criteria and likelihood

Security criteria	Likelihood
Confidentiality	If likelihood is higher than low, e.g., an unauthorized individual gets access to sensitive data of a crucial network or information system, which causes cyber threat risk incident, then immediately an incident response is required, mitigating or defending the unauthorized access in real-time
Integrity	If likelihood is higher than low, e.g., an unknown anomaly in network or information system behavior is identified, which causes a cyber threat risk incident, then immediately an incident response is required to reveal in real-time before the identified incident gets executed
Availability	If likelihood is higher than low means that a potential cyber threat risk incident can cause a • Denial of Services (DoS) attack to a crucial network or information system • Malfunction of data of a crucial network or information system If likelihood is higher than moderate, e.g., a service is unavailable for a period of time that happens not more than for example once for every month, then the incident behavior should be monitored for further decision

- What kind of malicious-damage, technical- and/or financial-wise has the cyber threat risk incident caused in industrial organizations?

Based on this background, a classification scheme can be derived for cybersecurity risk levels, in accordance with the acceptance instance in the CIA Triad. These distinct cybersecurity risk levels based on scores such as extreme, high, moderate, medium and low, as shown in Sect. 1.3.2. A methodological approach that use qualitative values for likelihood levels, described as frequency values, exploits cyber threat risk incidents. However, cyber threat risk incidents in organizations have different vulnerability types like:

- Internal, external and remote cyber threat risk incidents,
- Intrinsic vulnerability of network and information systems and infrastructure resources,
- Magnitude of hazards when networks and information systems and infrastructure resources cybersecurity is compromised,
- Cyber threat risk incidents actor's high motivation,
- And others.

Vulnerabilities are weaknesses enabling a cyber threat risk incident actor to denial, disturb or disrupt transmission of digital data required for undisturbed system operation. Therefore, identifying cyber threat risk incidents is an essential topic because they have increased in frequency and sophistication, presenting significant challenges for organizations that must defend their valuable data, assets and network and information systems from capable cyber threat actors. These actors range from individual, autonomous attackers to well-resourced groups operating in a coordinated manner as part of a cybercriminal organization (or on behalf of a nation-state). Cyber threat actors can be persistent, motivated, and agile, and use a variety of Tactics, Techniques, and Procedures (TTPs) to compromise networks, information systems, infrastructure resources and others, to disrupt services, commit financial fraud, and expose or steal intellectual property and other sensitive information. Thus, TTPs describe the behavior of a cybercriminal actor. Tactics are high-level descriptions of cybercriminal actor's behavior, techniques are detailed descriptions of behavior in the context of a tactic, and procedures are even lower-level, highly detailed descriptions in the context of a technique. TTPs could describe a cybercriminal actor's tendency to use a specific malware variant, order of operations, cyber-attack tool, delivery mechanism (e.g., phishing or watering hole attack), or exploit. Therefore, TTPs in cybersecurity describe how cybercriminals plan and execute their attacks. Thus, each component of TTP offers insight into different layers of cyber-attacks, enabling organizations cybersecurity teams understand and counteract potential cyber threat risk incidents. Given the risks cyber threat risk incidents present, it is increasingly important that organizations share Cyber Threat Intelligence (CTI) and use it to improve their cybersecurity posture.

CTI is any information about cyber threat risk incidents an organization has or is exposed to. CTI enable an organization to identify, assess, monitor, and respond to cyber threat risk incidents. Examples of CTI include indicators (system artifacts or observables associated with a cyber threat risk incident attack), TTPs, cybersecurity

1.6 Cybersecurity Is Paramount

alerts, cyber threat information reports, and recommended cybersecurity tool configurations. Most organizations already produce multiple types of cyber threat information that are available to share internally as part of their information technology and cybersecurity operations efforts [69].

However, the source of cyber threat risk incidents remains at the same time. It's always someone with an intension, which is the real source of cyber threat risk incidents. Hence, preventing and defending cyber threat risk incidents require Cyber Threat Intelligence (CTI), a knowledge enabling to prevent or mitigate them in an increasingly complex cyber threat landscape by monitoring and assessing potential cyber threat risk incident, determining system vulnerability, provide information on cybercriminal adversaries and ensuring the network and/or information system is secure. It also enables organizations to identify, prepare, and prevent cyber-attacks by providing information on cybercriminal attackers, their motivation, and capabilities. Furthermore, CTI represent information gathered from a variety of sources about current or potential cyber threat risk incidents against organizations.

Given that fact, cybersecurity is paramount, because cyber resilience is a nondebatable requirement for all organizations, which refers to an organization's ability to protect sensitive data, networks, information systems, infrastructure resources and others from cyber threat risk incidents, as well as to resume organizations business operations quickly in case of a successful cyber threat risk incident. This means the organization has integrated a comprehensive strategy to address cyber threat risk incidents effectively [70]:

- **Prevention:** Fundamental component of cyber resilience through preventive measures to reduce the likelihood of successful cyber treat risk incidents. Key practices include:
  - **Access Controls:** Implementing strict access controls to limit the exposure of sensitive data and network and information systems to unauthorized individuals,
  - **Cybersecurity Policies:** Developing and enforcing cybersecurity policies and procedures to maintain a cybersecure organizational environment,
  - **Employee Training:** Educating employees about cybersecurity awareness best practices, including how to identify phishing attempts and report suspicious activities,
  - **Regular Software Updates:** Keeping software, operating systems, and applications up to date to patch known vulnerabilities.

- **Detection:** Vital is early detection of cyber threat risk incidents to minimize their impact. Organizations should implement technologies and processes that enable swiftly identification of anomalies and potential security breaches. Key practices include:
  - **Cyber Intelligence (CI):** Information about emerging cyber threat risk incidents and vulnerabilities to proactively adapt cybersecurity measures and customize organizational plan to address likely and potentially impactful cyber threat risk incidents,

- **Intrusion Detection Systems (IDS):** Deploy IDS to monitor network traffic for suspicious activity,
- **Security Information and Event Management (SIEM) Systems:** Combines both Security Information Management (SIM) and Security Event Management (SEM) into one cybersecurity management system. SIEM technology collects event log data from a range of sources, identifies activity that deviates from the norm with real-time analysis, and takes appropriate action to cybersecurity. For this purpose, SIEM collect, aggregate, and analyze volumes of data from organization's applications, devices, servers, and users in real-time so cybersecurity teams can detect and block cyber threat risk incidents. SIEM tools offer many benefits that enable organizations overall cybersecurity posture, including:

  Advanced CI (ACI): Process that focuses on understanding the motivations, capabilities, and tactics of cyber threat incident actors, rather than just the individual cyber threat risk incidents themselves.
  Central view of potential cyber threat risk incidents,
  Greater transparency monitoring users, applications, and devices,
  Real-time cyber threat risk incident identification and response.
  Regulatory compliance auditing and reporting.
  Greater transparency monitoring users, applications, and devices

- **Response**: In the event of a cyber threat risk incident, a well-defined response plan must be in place within the organization, which enable a significant success in mitigating the potential damage. Key practices that covers the various aspects of handling a cyber breach and include:
  - **Communication:** Establishing communication protocols and channels to coordinate efforts and inform stakeholders.
  - **Containment:** Taking immediate action to contain the cyber breach and eliminate the cybercriminal attacker's presence.
  - **Legal and Regulatory Compliance:** Ensuring compliance with legal and regulatory requirements, including data breach notification laws like NIS2.
  - **Roles and Responsibilities:** Clearly define the roles and responsibilities of individuals involved in the response effort.
- **Recovery:** Cyber resilience goes beyond surviving a cyber threat risk incident; it includes the ability to recover and return to normal operations swiftly. Key practices of the recovery phase include:
  - **Data Backup and Restoration:** Regularly back up of critical data, network and information systems to enable restoration in the event of data loss through successful cyber threat risk incidents. Requires to determine the Recovery Point Objective (RPO) and Recovery Time Objective (RTO),
  - **Post-Incident Analysis:** Conduct a thorough post-incident analysis to identify weaknesses and improve future resilience efforts,

- **System Rebuild:** Rebuild compromised systems and infrastructure resources with improved cybersecurity measures.

Against this background, proactive measures are also important, for example, Secure by Design and Default (SbD). Secure-by-Design means that digital equipment built in a way that reasonably protects against malicious cybercriminals successfully gaining access to devices, data, infrastructure, and others. Secure by Default means digital equipment's are resilient against prevalent exploitation techniques out of the box without additional charge. With this in mind, implementing cyber resilience includes a cyber risk-focused plan that assumes the organizations will at some point face cyber threat risk incidents. Given that fact, the cyber resilience plan is based on four pillars:

- **Anticipate:** Organizations need to anticipate several types of cyber threat risk incidents, including
  - Cyber-attacks like:

    Ransomware: Persistent cyber threat risk incident, as attackers adapt their tactics to bypass solutions. Thus, type of malware attack in which attackers' locks and encrypts attacked data, important files and then demands a payment (ransom) to unlock and decrypt the data.

    Denial-of-Service (DoS) Attacks: Type of attack in which a malicious actor aims to render a computer or other device unavailable to its intended users by interrupting the device's normal functioning by flooding services or crashing services.

    Advanced Persistent Threat (APT): Describe an attack campaign in which an attacker, or team of attackers, establishes an illicit, long-term presence on a network in order to mine highly sensitive data.

    And others.

  - **Social-Engineering Attack**s: Psychological manipulation of people into performing actions or divulging confidential information. They show different types like:

    Phishing: Exploit human error to harvest credentials or spread malware, usually via infected email attachments or links to malicious websites.

    Spear-Phishing: Targeted version of a phishing scam whereby cybercriminals try to make their cyber-attack less conspicuous. It is much harder to detect and has better success rates if skillfully.

    And others.

- **Withstand:** Organizations need to withstand, which means ensuring that essential functions like Business Continuity (BC) can continue in case of cyber threat risk incidents, which requires identifying essential functions, along with all supporting processes, systems, services and infrastructure. Thus, steps to minimize the risk of those functions being disrupted by the cyber threat risk incident types identified.

- **Recover:** Organizations must restore essential functions during and after cyber threat risk incidents, which needs to prioritize recovery operations as well as considering usage of a phased approach. For example, restoring the most important Active Directory domain controllers can quickly get the business up and limping, if not running at full speed. However, it is necessary to trust that restoring a system will not restore the cyber threat risk incident such malicious software like Trojans or a backdoor for cybercriminals to regain access to the information systems.
- **Adapt:** Organizations need to adapt to the ever-evolving cyber threat risk incidents landscape, which never standstill, which require to regularly access the inventory of critical and crucial business functions and their supporting capabilities, as well as mitigation, response and restoration strategies. Cyber resilience is not a once-and-done event; it's a never-ending process in the digital age.

Advancements in the defensive capabilities and priority areas critically depend on progress in human aspects, research infrastructure, risk management, scientific foundations, and transition to practice. Thus, enhancing the trustworthiness of network and information systems, infrastructure resources and other is paramount, which requires enhancing creation and deployment practices of cybersecurity. Promising research involves AI and ML to detect errors in code, identify cybersecurity vulnerabilities, and make it easier for software engineers achieving cybersecurity by design. Therefore, to be better prepared for cybersecurity purposes, governments worked out cybersecurity directives.

In the European Union (EU) the directive on Network and Information Security (NIS) has been introduced to regulate cybersecurity objectives, and to achieve a high common level of cybersecurity across the EU-Member States. The background of NIS is the growing cybersecurity demands in the era of digital transformation with its manifold of digital-based digital resources. The further development of NIS result in NIS2 (Chap. 2), a cybersecurity-strategy with the focus on the directive of Critical Entities Resilience (CER), a network of Security Operations Centers (SOCs) and new measures, because the cyber threat risk incidents landscape has changed considerably. In this sense, the NIS2 cybersecurity directive refers to:

- Reinforced rules for a high common level of cybersecurity across the EU,
- Supervision for medium and large organizations across more than a dozen key sectors,
- Establishment of a higher level of mandatory, reviewable and sanctionable cybersecurity measures for cyber risk management, security governance, incident reporting/recovery, resilience and network, system and application security,
- Risk reviews of cybersecurity practices for major connected third-party services providers.

Currently, the NIS2 directive is prepared for official publishing by the EU parliament, which happen in 2024. Thereafter, adoption by national legislative bodies across the EU Member states is done that NIS2 can come into binding effect no later than October 2024.

1.6 Cybersecurity Is Paramount

In 2014, the Executive Office of the President of the United States has published the Cybersecurity Enhancement Act (Public Law 113–274) that requires the National Science and Technology Council and the Networking and Information Technology Research and Development Program to develop, maintain, and update every 4 years a Cybersecurity Research and Development (R&D) Strategic Plan. This plan also addresses priorities established by the National Cyber Strategy of the United States or America, including both its domestic and foreign priorities, and by the Administrations FY 2021 Research and Development Budget Priorities Memorandum. The Plan identifies the following goals for cybersecurity R&D [71]:

- Understand human aspects of cybersecurity,
- Provide effective and efficient risk management,
- Develop effective and efficient methods for deterring and countering malicious cyber-activities,
- Develop integrated safety-security framework and methodologies,
- Improve systems development and operation for sustainable cybersecurity

To realize the goal of a cybersecure cyberspace, the Plan carries forward the essential concepts from the 2024 Federal Cybersecurity Research and Development Strategic Plan [72], including the framework of four interdependent defensive capabilities [72]:

- **Deter:** Ability to efficiently discourage malicious cyber activities by increasing costs, cyber threat risk incidents, and uncertainty to adversaries and diminishing their spoils.
- **Protect:** Ability of components, systems, users, and critical infrastructure resources to efficiently resist malicious cyber activities and to ensure confidentiality, integrity, availability (CIA Triad), and accountability.
- **Detect:** Ability to efficiently detect, and even anticipate, adversary decisions and activities, given that systems should be assumed to be vulnerable to malicious cyber activities.
- **Respond:** Ability to dynamically react to malicious cyber activities by adapting to disruption, countering the malicious activities, recovering from damage, maintaining operations while completing restoration, and adjusting to be able to thwart similar future activities.

1.6.1 NIST Cybersecurity Framework

The four foregoing mentioned elements are similar but not identical to the five core functions in the National Institute of Standards and Technology's (NIST) Cybersecurity Framework (CSF) for improving critical infrastructure cybersecurity (see Sect. 3.2.1), which includes the following components:

- **NIST CSF Core:** Nucleus of the NIST CSF, which is a taxonomy of high-level cybersecurity outcomes that can help any organization manage its cybersecurity risks. NIST CSF Core components are a hierarchy of Functions, Categories, and Subcategories that detail each outcome. Outcomes can be understood by a broad audience, including executives, managers, and practitioners, regardless of their cybersecurity expertise. Because the outcomes are sector-, country-, and technology-neutral, they provide an organization with the flexibility needed to address its unique risks, technologies, and mission considerations.
- **NIST CSF Organizational Profiles:** Mechanism for describing an organization's current and/or target cybersecurity posture in terms of the NIST CSF Core's outcomes.
- **NIST CSF Tiers:** Applied to NIST CSF Organizational Profiles to characterize the rigor of an organization's cybersecurity risk governance and management practices. Tiers can also provide context for how an organization views cybersecurity risks and the processes in place to manage those risks.

An organization can use the NIST CSF Core, Profiles, and Tiers with the supplementary resources [73] to:

- **Understand and Assess:** Describe the current or target cybersecurity posture of part or all of an organization, determine gaps, and assess progress toward addressing those gaps.
- **Prioritize:** Identify, organize, and prioritize actions for managing cybersecurity risks that align with the organization's mission, legal and regulatory requirements, and risk management and governance expectations.
- **Communicate:** Provide a common language for communicating inside and outside the organization about cybersecurity risks, capabilities, needs, and expectations cybersecurity risks.

The NIST CSF is designed to be used by organizations of all sizes and sectors, including industry, government, academia, and nonprofit organizations, regardless of the maturity level of their cybersecurity programs. The NIST CSF is a foundational resource that may be adopted voluntarily and through governmental policies and mandates. The NIST CSF's taxonomy and referenced standards, guidelines, and practices are not country-specific, and previous versions of the NIST CSF have been leveraged successfully by many governments and other organizations both inside and outside of the United States. Compliance with the NIST CSF is not mandatory for all organizations. The decision to comply with NIST CSF depends on various factors and is influenced by the type of organization and its relationship with the government. For federal agencies, compliance with the NIST CSF is mandatory.

The Federal Cybersecurity Research and Development Strategic Plan [72] also advance one or more of the following Priority Areas defined in the Plan:

- **Artificial Intelligence (AI):** Capabilities that enable computers and automated and other digital systems to perform tasks that have historically required human cognition and what are typically considered human decision-making abilities.
- **Quantum Information Science (QIS):** Capabilities that harness quantum mechanics and quantum material properties to achieve computation, information

processing, communications, and sensing in ways that cannot be achieved by classical physics principles.

- **Trustworthy Distributed Digital Infrastructure (TDDI):** Technologies that facilitate secure information communications infrastructure, enabling next-generation wireless communication, distributed computing, seamless integration of telecommunication systems with Cyber-Physical Systems (CPS), and provides the communications infrastructure for the Industries of the Future. CPS are embedded systems of integrating digitized process systems with digital communication primarily developed to monitor and control the physical devices in the systems. IoT and IIoT are the key technologies to network the physical process system.
- **Privacy:** Solutions that minimize privacy risks or prevent privacy violations arising from the collection and use of people's private information.
- **Cybersecure Hardware and Software (HW & SW):** Technologies that provide and improve security properties of hardware and software components in computing and communication systems.
- **Education and Workforce Development:** Programs in cybersecurity education, training, and professional development to sustain cybersecurity innovations by the national workforce.

Therefore, advancements in the defensive capabilities and priority areas critically depend on progress in human aspects, research infrastructure, risk management, scientific foundations, and transition to practice. Hence, enhancing the trustworthiness of computer systems, network and information systems, infrastructure resources and others is paramount, which requires enhancing creation and deployment practices of cybersecurity. Promising research involves AI and Machine Learning (ML) to detect errors and/or insecure code in programs, identify cybersecurity vulnerabilities and others, and make it easier for software engineers achieving cybersecurity by design in their developed crucial and critical systems. This requires AI-powered capabilities to analyze large amounts of data from applications, devices and networks, based on which AI continuously develop more advanced cybersecurity measures to update existing measures. At its end this enables a dynamic cybersecurity approach for an ever-changing cyber threat risk event landscape in the direction of a Zero Trust environment Zero Trust is a cybersecurity model that requires all users, whether in or outside an organization's network, be authenticated, authorized, and continuously validated for cybersecurity configuration and posture, before being granted or keeping access to organizations applications, devices or networks.

1.6.2 Artificial Intelligence in Cybersecurity

Artificial Intelligence (AI) is entirely within computational theory, which intersects with several other disciplines, including mathematics, cognitive science, neuroscience, computer science, and others. In AI, subsets are applied, whereby subsets refer to a specialized area of AI. Subsets of AI include Machine Learning (ML),

Natural Language Processing (NLP), speech recognition, expert systems, and other. In this context, subsets focus on specific applications of AI, each utilizing various techniques, methodologies and models, to address particular task or challenges within the field. In its broadest definition AI is equated with algorithms, and in its strictest definition, from the Oxford Dictionary of Phrase and Fable: "The theory and development of computer systems able to perform tasks normally requiring human intelligence, such as visual perception, speech recognition, decision-making, and translation between languages." Against this background, in an organization an AI strategy contains a comprehensive set of policies, standards, roles, processes, tools, training, and others to ensure that AI systems are deployed and used in a responsible manner. Furthermore, AI based learning can be defined as innovative method with the capability to learn solving challenging problems, applying computer algorithms and advanced subsets ranging from cognition, decision-making, speech recognition, pattern recognition, and other elements. Hence, AI is a broad concept enabling a machine or digital system sensing, reasoning, acting, or adapting similar to a human. For this purpose, AI applications result in different practices like:

- **Adaptive AI:** Practice that support a decision-making framework centered around making faster decisions while remaining flexible to adjust as issues arise. These systems aim to continuously learn based on new data at runtime to adapt more quickly to changes in real-world circumstances. The AI engineering framework can help orchestrate and optimize applications to adapt to, resist or absorb disruptions, facilitating the management of adaptive systems [74].
- **Agentic AI:** Practice that leverages an AI agent style of software implementation where AI agents are autonomous or semiautonomous software entities that use AI techniques to perceive, make decisions, take actions and achieve goals in their digital or physical environments [75].
- **Composite AI:** Practice that refers to the combined application (or fusion) of different AI techniques to improve the efficiency of learning to broaden the level of knowledge representations. Composite AI broadens AI abstraction mechanisms and, ultimately, provides a platform to solve a wider range of business problems in a more effective manner [76].
- **Generative AI (GenAI):** Practice that can learn from existing artifacts to generate new, realistic artifacts (at scale) that reflect the characteristics of the training data but don't repeat it. It can produce a variety of novel content, such as images, video, music, speech, text, software code and product designs, but they are, in essence, prediction algorithms [77], and captured the imagination of researchers alike, opening up new avenues for creativity. However, GenAI has also emerged as attractive tool for cybercriminals to increase the efficacy of cyber-attacks particularly those using social engineering efforts with convincing outputs without precise prompting, custom model training, or fine tuning [78]. Furthermore, employing GenAI to Large Language Models (LLMs) especially enable to prevent harm by identifying potential cyber risk sources and addressing them before they can negatively impact, e.g., in regard to spam email campaigns.
- **And others.**

1.6 Cybersecurity Is Paramount

With it, AI as transformative technology changing the landscape of organizations for almost every sector. Thus, applying AI need casting a vision and defining clear outcomes for any AI project, which is key achieving goals and ensuring a successful project outcome. In this sense, AI is also umbrella of Machine Learning (ML) and Deep Learning (DL).

ML algorithms are trained on labelled or unlabeled data to extract knowledge from data and learn from it, to make predictions or decisions, enabling automation and efficiency in various tasks. In contrast, Deep Learning (DL), a subset of AI, uses multiple-layer Artificial Neural Networks (ANN) to learn with available data and make decisions or predictions. ANNs are artificial adaptive systems in-spired by a model of processes of the human brain. Hence, ANNs are systems that allow modifications of their internal structure in relation to a functional objective. They are particularly suited for solving nonlinear problems. The basic elements of ANNs are nodes, also called Processing Elements (PE), and the connections. Each node has its own input, from which it receives information from other nodes and/or from the environment and its own output, from which it communicates with other nodes or with the environment. Finally, each node has a transfer function through which it transforms its own global input into output. The connections between the nodes can modify themselves. The way through which the nodes modify themselves is called the learning law. The learning process is one key mechanism that characterize ANNs as adaptive processing systems. The node structure of an ANN is shown in Fig. 1.2 [79].

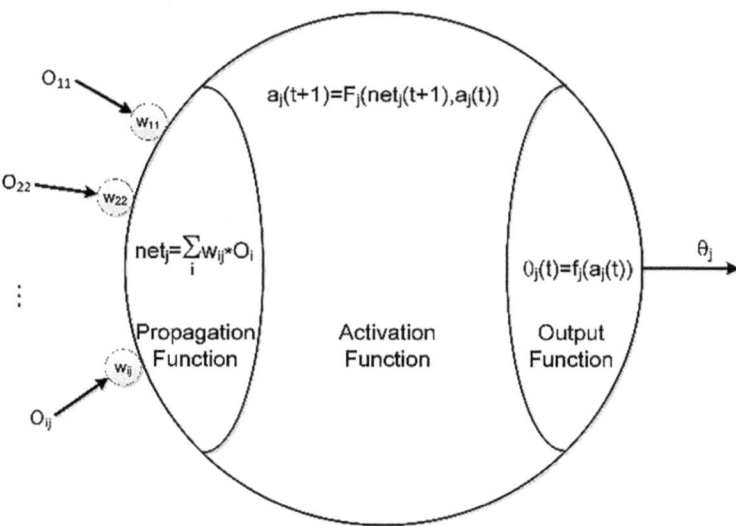

Fig. 1.2 ANN node structure

ChatGPT

ChatGPT is an AI chatbot that uses Natural Language Processing (NLP) to create humanlike conversational dialogue such as text generation, question answering, context extraction, and other. In this context, ChatGPT uses, e.g., Deep Learning (DL), a subset of ML, to produce humanlike text through transformer neural networks. The Large Language Model (LLM) respond to questions and compose various written content and providing explanations.

ChatGPT works through its Generative Pre-trained Transformer (GPT), which uses specialized algorithms to find patterns within data sequences. Originally ChatGPT used a Large Language Model (LLM), a neural network machine learning model and in GPT-3 the third generation of GPT. The transformer pulls from a significant amount of data to formulate a response.

In order for the model to comprehend the context and meaning of natural language text, it is trained on a huge database of text. Against this background, ChatGPT intelligence is to understand relationships and patterns from the input text and generate or predict new text that is homogeneous to the input/training data. With it, ChatGPT is able to respond to questions and prompts in a manner that is comparable to that of a human, making it useful in a wide scope by using an LLM. An LLM is trained with a high number of examples to be able to recognize and interpret human speech or other types of complex data. Many LLM are also trained on data from the Internet. The LLM is required to determine probabilities for prompts. Hence, when the LLM generates a function call in its response, it adds a new field function_call that contains the name of the function that the LLM wants to call and the values of the arguments for that function. If the LLM decides not to make a function call, then the function_call field in the response is not set. By checking if the function_call field is set in the response or not, is known, when the LLM make a function call. In this regard, ChatGPT is based on generative language models of the GPT family. These are statistical models that assign a probability to a sequence of words, fundamental to many NLP tasks [80].

Machine Learning in Cybersecurity

AI plays an important role in many applications today. A specific application of AI and especially it's subset ML is on the cybersecurity side to defend cyber threat risk incidents like fraud detection. In this case, ML models analyze transaction patterns to detect anomalies and prevent frauds in real-time. Therefore, ML techniques used by cybercriminals search for organizations vulnerabilities to apply their sophisticated cyber threat risk incidents ever faster and thus overcoming the defensive wall of an organization ever more easily. Therefore, ML for organizations cybersecurity must enable to detect and combat cyber threat risk incidents faster and more likely.

On the defense side, ML techniques play an important role in providing robust and intelligent techniques to improve performance in detecting cyber threat risk incidents swiftly. Given this fact, a three-step approach is used to investigate,

1.6 Cybersecurity Is Paramount

identify and report about the findings of cyber threat risk incidents. Investigate means representing information in conjunction with some kind of reasoning in order to do inference. Identifying refers to sorting the many log records to find out where are anomalous activities and/or where are the outliers. After a successful execution of these steps, a report is essential to close the cybersecurity analysis cycle, complying with regulation requirements. With it, the three-step sequence enables to find out in almost less time what is the signature or pattern of the potentially intruded cyber threat risk incident that interfere with regular system operation, based on which one can act to start with cyber-defense against cybercriminal attacks, for example identified malicious code, malware, phishing and others. With this in mind, AI is increasingly important for prevention against cybersecurity risks through cyber threat risk incidents. Unfortunately, detection of cyber threat risk incidents is not simple, because cybercriminals can modify their cyber threat risk incidents to evade detection and increasingly using AI technologies to make their cyber threat risk incidents more rigor in order to achieve negative consequences to organization. In this regard, cyber threat risk incidents on vulnerabilities be optimized using ML or automated Social Engineering Attacks, which is a serious problem for organizations cybersecurity, as AI-based exploited cyber-attacks can become difficult to detect and combat. Cybercriminal abuse of generative AI raises concerns about social engineering campaigns and the creation of malicious software, tools, and resources to conduct dangerous attacks. Trends already prove AI was often used for social engineering, and the power of AI creates endless possibilities for adversaries to become even more sophisticated.

ML as subset of AI enabling computers to learn from digital data and make decisions or predictions without being explicitly programmed to do so. At its core, ML is creating and implementing algorithms that facilitate decisions and predictions to applications like detect and defend against cyber threat risk incidents in a variety of ways. For this purpose, ML can detect cyber threat risk incidents by constantly monitoring the behavior of network and information systems for anomalies. Moreover, ML can detect malware in encrypted traffic in networks, analyzing encrypted traffic elements in the network. Furthermore, ML can pinpoint malicious patterns to find cyber threat risk incidents hidden with encryption, and others. In this context, all ML algorithms be trained to detect anomalous behavior patterns that indicate probable cyber risks, and block similar malware files in real-time, or trained to identify emails that contain phishing attacks by looking for suspicious content, URLs or sender addresses. ML models also enable being used to identify cyber risk vectors to detect vulnerabilities in network and information systems, which can be exploited by cybercriminals, and thereafter close them. Therefore, AI-powered monitoring by ML is used to detect, for example, suspicious anomalies, and react accordingly to make crucial and critical network und information systems, infrastructure resources and others cyber-secure. However, types of anomalies can vary by organization and business. Therefore, anomaly detection means defining regular (normal) patterns and metrics, based on regular business functions and goals, and identifying digital data points that fall outside of regular (normal) organizations operation's behavior, for example, higher than average digital data traffic on a website or traffic for a particular period of time, which can be assumed as pattern of a cyber threat risk incident, or detect never-before-seen malware that is trying to run on endpoints. In such cases a detection system is needed that automatically trigger cyber threat risk incident detection alerts. However, anomalies are not inherently bad, but organizations must be aware of them, and need digital data to put them in context, which is essential to understand and protect its business.

In this context, ML techniques are leveraged to detect anomalous behavior through different methods [10]:

- **Supervised Machine Learning (SML):** Category of ML that use labeled datasets to train the ML algorithm to predict outcomes that are either regular (normal) or abnormal in contrast to used digital training data. The SML model is trained with labeled training datasets to learn the relationship between the input and output, to enable detecting outliers based on the examples it is given, and thus useful in known outlier detection. However, SML is not capable of discovering unknown anomalies or predicting future issues. Common algorithms used in SML are referred to in Table 1.14:
- **Unsupervised Machine Learning (UML):** Do not require labeled data-sets and can handle more complex datasets. Common types of unsupervised learning are clustering, association rules, and dimensionality reduction. These methods can find patterns from input data and make assumptions about what dataset is perceived as regular (normal). Common algorithms used in UML are referred to in Table 1.15.
- **Semi-Supervised Machine Learning (SSML):** Method combines the benefits of the previous methods. It can be applied to automate feature learning and work with unstructured datasets. However, by combining it with human supervision, an opportunity occurs to monitor and control what kind of patterns the model learns. This usually enables to make the model's predictions more accurate.

Table 1.14 SML algorithm with capabilities

SML algorithm	Capability
Linear Regression	Regression method used for anomaly detection. It is a statistical tool used to find the relationship between labeled data and variable data, based on the assumption that similar data points will be found near each other. If a data point appears further away from a dense section of points, it is considered an anomaly
Logistic Regression	Method to predict a binary outcome (yes, no; 1,0) based on one or more independent variables. It is an algorithm for solving classification problems
Naïve Bayes Algorithm	Method to predict the probability of an event based on prior knowledge. It's an algorithm for solving classification problems
K-nearest Neighbor	Method to find the K nearest neighbors of a data point. It's an algorithm for solving classification and regression problems
And others	…

1.6 Cybersecurity Is Paramount

Table 1.15 UML algorithm with capabilities

UML algorithm	Capability
K-means Algorithm	Partitioning data into K clusters represented by their centers of means. Algorithm refers to points in the cluster center to find locally optimal solutions to the clustering error. These clusters be used to find patterns and make inferences about data that is found to be out of the ordinary
Isolation Forest Algorithm	Create decision trees that map out data points and randomly select an area to analyze. Process is repeated, and each point receives an anomaly score between 0 and 1, based on its location to the other points. Values below .5 are generally considered to be regular, while values that exceed that threshold are more likely to be anomalous
Autoencoder	Method is a particular type of feed-forward network, used to decompress and compress input data. Decompression and compression are data-specific, which means that the autoencoder only be able to actually compress data on which it has been trained, and give a bad performance for untrained (anomalous)
Deep Belief Network	Method based in Deep Learning (DL), a subset of ML. It creates a hierarchy of layers. The first layer consists of the input data, the last layer consists of neurons used for classification. Training is fast since training only occurs in one direction, fom input to output
And others	…

Against this background, all AI based ML methods use specific mathematical algorithms to create ML-models for predictions on input dataset to enable detecting and defending malicious cyber activities in near real-time. For this purpose, ML algorithms offers several benefits in cybersecurity applications, including:

- **Faster Detection of Cyber Threat Risk Incident Attacks:** ML enable to detect anomalies in real-time and alert faster response time the cybersecurity personnel to act.
- **Greater Accuracy:** ML enable to process large amounts of datasets and recognize complex patterns in the datasets, which lead to greater accuracy in detecting cyber threat risk incidents.
- **Automated Defenses:** ML enable automatically defenses, e.g., blocking suspicious network traffic, capsuling infected endpoints, and others.

As illustrated in Tables 1.14 and 1.15, a variety of ML algorithms exist that are suitable for different tasks and/or applications. These include neural networks, decision trees, random forests, support vector machines, cluster analyses, and others. The choice of appropriate learning algorithms depends on the type of datasets to be processed and the specific requirements and goals of the ML cybersecurity application and can be assigned as described in [81]:

- **SML in cybersecurity application include:**
  - Identifying unique labels of network risks, such as scanning and spoofing by using linear regression methods,

- Predicting or classifying a target variable for a specific cyber threat risk incident like Distributed Denial of Service (DDoS) attack,
  - Training models on benign and malicious samples to help them predict whether new samples are malicious, and others.
- **UML in cybersecurity application include:**
  - Detecting unusual behavior based on decision tree methods,
  - Identifying new attack patterns,
  - Mitigating zero-day attacks.
  - And others.
- **SSML in cybersecurity application include:**
  - Adversarial neural networks,
  - Malicious and benign bot identification,
  - Malware detection based on k-means clustering,
  - Ransomware detection,
- And others

The selection and possible adaptability of ML algorithms influences how successfully to be trained and optimized in order to make precise predictions or decisions. With regard to the database used, it is important that the dataset is of high quality, sufficient and representative of the problem to be solved with ML algorithms. With this in mind, it is essential to continuously monitor and validate the performance of ML algorithm used to ensure that it consistently ensure optimized predictions and decisions that meet organizations expectations. This is an ongoing procedure that must be carried out throughout the entire lifecycle of using the ML algorithm, including at least the following areas:

- **Training Validation:** ML performance that include accuracy, predictive ability and error rate, which must be checked during the training phase.
- **Test Data Validation:** Optimized test data used must be checked for independency, to respond effectively to new datasets to make precise predictions.
- **Performance Monitoring:** ML performance must be continuously monitored in real-time.
- **Data Quality Monitoring:** Datasets used to be processed must always be correct and require continuous monitoring of dataset quality and integrity.

With this in mind, ML-based models detect patterns in input datasets that be compared with measured datasets, predicting whether these are applicable as expected or not, whereby the latter case indicating that cyber threat risk incidents are present. The identified cyber threat risk incident can refer in a negative impact on the organizations regular operational state, which result in a certain degree of uncertainty. Hence, ML analyze detected potential cyber threat risk incidents to respond on them immediately in an automated way. Therefore, ML is a method to deal

1.6 Cybersecurity Is Paramount

with uncertainty to gain a clear decision making. However, an effective ML strategy is needed that requires planning based on four properties, as reported by Gartner [82]:

- **Vision:** ML drive organization goals and expected benefits through measure success by a success metrics.
- **Value:** ML remove hurdles that hinder success through business impact, change management, people and skills.
- **Risks:** ML assess and mitigate organizations cyber threat incident risks through regulatory, reputational, competency and technology.
- **Adoption:** ML pursue, based on organizations value and feasibility by using cases and value maps, ML decision framework, and decision governance.

Anomaly Detection, Misuse Detection and Other Approaches

AI and its subsets ML and DL use cybersecurity measures, detecting and preventing cyber threat risk incidents on digital data or assets and be broadly assigned to misuse and/or anomaly detection.

In misuse detection, each misuse class is learned by using appropriate scenarios from training digital data sets. During test, digital data or patterns gathered from the object under investigation, and thereafter classified whether they belong to one of the misuse classes or not. In case digital data or patterns does not belong to any of the misuse classes from the training set, it is classified as regular.

In anomaly detection regular patterns are defined during training, and thereafter learned patterns compared to pattern, obtained from the object under test. Finally, every inacceptable measured pattern is classified as anomalous in correlation to the regular pattern. However, if the problem of anomaly detection is ill-posed, it's hard to decide what an anomaly is.

In this sense, cybersecurity intrusion detection is the continuous process monitoring for harmful cyber threat risk incidents that occur, by means of analyzed signs of probably cyber threat risk incidents, to interdict unauthorized access [31]. This is typically accomplished by collecting digital data from a variety of sources, and thereafter analyzing them for presumably cyber threat risk incidents by means of identifying unusual digital data. Consequently, the main task of a ML based Intrusion Detection System Architecture (IDSA) is detect digital data, patterns or signatures of probable cybersecurity breaches as result of cyber threat risk incidents. Therefore, IDSA design and application aimed to deter harmful damage or mitigate damage caused by cybercriminals [31].

Another approach of AI based intrusion detection combines the outputs of classification methods as collective output that enhance classification performance. In this sense, ensemble learning enables combining heterogeneous and homogeneous multi-classifiers to obtain sustainable classification results [37, 38]. Thereby, the ensemble learning use different ML methods to reduce variance and become robust

Table 1.16 ML models for cybersecure organizations digital data, assets, infrastructure resources and others

ML model	Working principle	Advantages	Disadvantages	Application(s)
Associated rule-based learning method	Investigates relationships between variables in digital training data sets	Simple algorithms, easy to use	Execution time of algorithm is high due to complexity	Intrusion Detection, Cyber risk detection and protection
Ensemble learning method	Combines concepts of different classification methods	Robust algorithm, adapt classifier-based methods compared to single classifier	Execution time of algorithm is high due to complexity	Intrusion Detection, Anomaly Detection, Malware Detection

to over-fitting [39]. Table 1.16 illustrate the potential of two AI subset ML methods, to cybersecure organization digital data, assets, infrastructure resources and others.

To summarize, AI's subsets DL and ML rely on intelligent algorithms, whereby DL focuses on feature extraction through its layered architecture, while ML often requires manual feature engineering. Moreover, DL models rely on managing massive volumes of data throughout the entire data pipeline including both the training and inferencing phase, compared to traditional ML models.

However, a new form of cybercrime in AI applications recently appear to GPT-4 OpenAI Service, Google Cloud Vertex AI, DeepSeek-R1, and other elements, the so-called Large Language Model (LLM) Jacking attacks. Cybercriminals use AI login credentials to use costly AI services without authorization, resulting in massive financial losses of the attacked organization [83]. Furthermore, GenAI, a type of AI that can create new content, trained by algorithms based on large data sets, has emerged as an attractive approach for cybercriminals with a low barrier to entry, which makes it applicable on a broad range. Recent advancements in genAI with specific LLLs have enhanced the efficacy of certain cyber-attack operations, particularly those using Social Engineering efforts or Malicious Computer Network Operations (CNO). CNO is used in civilian and military applications, to leverage and optimize computer networks to improve organizations capability, or to gain information on cyber attackers' malicious activity to impede use this enabling capability.

1.7 Exercises

- What is meant by the term *Digitalization?*
  - Describe the characteristics and capabilities of Digitalization
- What is meant by the term *Digital Transformation?*
  - Describe the characteristics and capabilities of Digital Transformation

1.7 Exercises

- What is meant by the term *Industry 4.0*?
  - Describe the efforts and potential application domains of Industry 4.0
- What is meant by the term *Cultural Transformation*?
  - Describe the characteristics and capabilities of the Cultural Transformation
- What is meant by the term *Domain Transformation*?
  - Describe the characteristics and capabilities of the Domains Transformation
- What is meant by the term *Process Transformation*?
  - Describe the characteristics and capabilities of the Process Transformation
- What is meant by the term *Emerging Technologies*?
  - Describe some emerging technologies and their capabilities applied to Digital Transformation
- What is meant by the term *Artificial Intelligence*?
  - Describe the characteristics and capabilities of Artificial Intelligence
- What is meant by the term *Big Data and Analytics*?
  - Describe the characteristics and capabilities of Big Data and Analytics
- What is meant by the term *Cloud Computing*?
  - Describe the characteristics and capabilities of Cloud Computing
- What is meant by the term *Cloud-as-a-Service*?
  - Describe the main four cloud service models in detail
- What is meant by the term *Digital Twin*?
  - Describe the characteristics and capabilities of a Digital Twin
- What is meant by the term *Industrial Internet of Thing*?
  - Describe the characteristics and capabilities of the Industrial Internet of Things
- What is meant by the term *Mobile Technology*?
  - Describe the characteristics and capabilities of Mobile Technology
- What is meant by the term *Cybersecurity*?
  - Describe the key characteristics and capabilities of Cybersecurity
- What is meant by the term *Network and Information Security Directive (NIS2)*?
  - Describe the key characteristics and capabilities of the Network and Information Security Directive (NIS2)
- What is meant by the term *National Institute of Standards and Technology Cybersecurity Framework 2.0 (NIST CSF 2.0)*?
  - Describe the key characteristics and capabilities of the National Institute of Standards and Technology Cybersecurity Framework 2.0 (NIST CSF 2.0)
- What is meant by the term *Cybersecurity Situational Awareness*?
  - Describe the major characteristics and capabilities of Cybersecurity Situational Awareness

- What is meant by the term *Cybersecurity Risk Assessment*?
  - Describe the major characteristics and capabilities of Cybersecurity Risk Assessment
- What is meant by the term *Cybersecurity Risk Management*?
  - Describe the major characteristics and capabilities of Cybersecurity Risk Management
- What is meant by the term *Cybersecurity Maturity Level Model*?
  - Describe the key characteristics and capabilities of Cybersecurity Maturity Level Model
- What is meant by the term *Radar Chart*?
  - Describe the major characteristics and capabilities of the Radar Chart
- What is meant by the term *NIST CSF Core*?
  - Describe the characteristics and capabilities of the NIST CSF Core
- What is meant by the term *NIST CSF Organizational Profile*?
  - Describe the major characteristics and capabilities of the NIST CSF Organizational Profile
- What is meant by the term *NIST CSF Tiers*?
  - Describe the major characteristics and capabilities of the NIST CSF Tiers.
- What is meant by the term *Cybersecurity Maturity Level Model?*
  - Describe the major characteristics and capabilities of Maturity Level Criteria
- What is meant by the term *Vulnerability*?
  - Describe the characteristics and capabilities of Vulnerability risks
- What is meant by the term *Compliance Risk*?
  - Describe the advantages and disadvantages of the Compliance Risk
- What is meant by the term *Operational Risk*?
  - Describe the advantages and disadvantages of Operational Risks
- What is meant by the term *Operational Technology?*
  - Describe the characteristics and capabilities of Operational Technology
- What is meant by the term *OT Security*?
  - Describe the major characteristics and capabilities of Operational Technology Security
- What is meant by the term *CIA Triad*?
  - Describe the major characteristics and capabilities of the C, I and A
- What is meant by the term *Artificial Intelligence*?
  - Describe the term Artificial Intelligence by the Oxford Dictionary of Phrases and Fable
- What is meant by the term *GenAI?*

- Describe the main characteristics of GenAI
- What is meant by the term *ChatGPT*?
  - Describe the major characteristics and capabilities of ChatGPT
- What is meant by the term *LLM Jacking Attack?*
  - Describe the main characteristics of LLM Jacking
- What is meant by the Term *Computer Network Operations (CNO?*
  - Describe the main differences between civilian and military usage
- What is meant by the term *Machine Learning*?
  - Describe the major characteristics and capabilities of Machine Learning
- What is meant by the term *Supervised Machine Learning*?
  - Describe the major characteristics and capabilities of Supervised Machine Learning
- What is meant by the term *Unsupervised Machine Learning*?
  - Describe the major characteristics and capabilities of Unsupervised Machine Learning
- What is meant by the term *Semi-Supervised Machine Learning*?
  - Describe the major characteristics and capabilities of Semi-Supervised Machine Learning.
- What is meant by the term *Semi-Supervised Machine Learning*?
  - Describe the major characteristics and capabilities of Semi-Supervised Machine Learning.

References

1. Hess, T., Benlian, C., Wiesbck, F.: Options for Formulating a Digital Transformation Strategy. In: MIS Q. Exec., Vol. 15, No. 2, pp.123139, 2016
2. Nadkarni,S. Prügl, R.: Digital Transformation: A Review, Synthesis and Opportunities for future Research. In: Management Review Quarterly, Vol. 71, pp. 233-341, 2021
3. Verhof, P.C., Broekhuizena, T., Barth, Y., Battachayaa, A., Donga, J.Q., Fabiana, N., Haenleine, M.: Digital Transformation: A Multidisciplinary Reflection and Research Agenda. In: Journal of Business Research, Vol. 122, pp. 889-901. 2021
4. Lasi, H., Fettke, P., Kemper, H.-G., Feld, T., Hoffmann, M.: Industry 4.0. In: Business and Information Systems Engineering, Vol. 6, No. 4, pp.239-242. 2014
5. Möller, D.P.F.: Guide to Computing Fundamentals in Cyber-Physical Systems: Concepts, Design Methods, and Applications. Springer International Publisher, 2016
6. Hund, A., Wagner, H.-T., Beimborn, D., Weitzel, T.: Digital Innovation: Review and Novel Perspective. In: Journal of Strategic Information Systems. Vol. 30, No. 4, 101695, 2021
7. Gökalp, M. O.; Kayabay, K., Akyol, M. A., Eren, P. E., Kocyigit, A.: Big Data for Industry 4.0: A Conceptual Framework. In: 2016 Intern. Conf. on Computational Science and Computational Intelligence, pp. 431-434, 2016

8. Sharma, A., Pandey, H.: Big Data and Analytics in Industry 4.0. In: Nayyar, A., Kumar, A. (Eds) A Roadmap to Industry 4.0: Smart Production, Sharp Business and Sustainable Development, pp.57-72, Springer Publ., 2020
9. Loshin, D.: Big Data Analytics: From Strategic Planning to Enterprise Integration with Tools, Techniques, NoSQL and Graph. Morgan Kaufman Publ. 2013
10. Möller, D. P. F.: Guide to Cybersecurity in Digital Transformation—Trends, Methods, Technologies, Applications and Best Pest Practices. Springer Nature. 2023
11. Chaouchi, H. /Ed.): The Internet of Things - Connecting Objects to the Web. Wiley and Sons, 2010
12. Nambisan, S., Wright, M., Feldman, M.: The Digital Transformation of Innovation and Entrepreneurship: Progress, Challenges and Key Themes. In: MIS Quarterly, Vol. 41, No.1, pp.223-238, 2017
13. Lazzazzara, A., Ricardi, F., Za, S.: Exploiting Digital Ecosystems: Organizational and Human Challenges, Springer Nature, 2020
14. Downes, L., Nunes, P.: Big Bang Disruption. In: Harvard Business Review Vol. 91, No. 3, pp. 44-56, 2013
15. Westerman, G., Calméjane, C., Bonnet, D., Ferraris, P., McAfee, A.: Digital Transformation: A Roadmap for Billion-Dollar Organizations. In: MIT Center for Digital Business and Capgemini Consulting, pp. 1-68, 2011
16. Fitzgerald, M., Kruschwitz, N., Bonnet, D., Welch, M.: Embracing Digital Technology: A New Strategic Imperative. IN: MIT Sloan Management Review Vol. 55, No. 2, pp. 1-12, 2014
17. Hartl, E., Hess, T.: Cultural Values in Digital Transformation: The Insight from a Delphi-Study. In: Proceed. 23rd Americas Conference on Information Systems, pp. 1-10, 2017
18. Bharadwaj, N., Noble, C.: Finding Innovations in Data Risk Environments. In: Journal in Production Innovation Management, Vol. 34, No.5, pp. 560-564, 2017
19. Grassmann, O., Sutter, P.: Digital Transformation (in German). Hanser Publ. 2013
20. Tolboom, I.: The Impact of Digital Transformation. MSc-Thesis, TU Delft, 2016
21. Schwertner, K.: Digital Transformation of Business. In. Trakia Journal of Sciences, Vol. 15, Suppl. 1, pp. 388-393, 2017. https://pdfs.semanticscholar.org/51bb/4fd609d174438fb8911f28 3d48d34ef1e894.pdf/1000 (accessed 02.2023)
22. Chanias, S., Hess, T.: Understanding Digital Transformation Strategy: Insights from Europe's Automotive Industry. In: PACIS Proceedings, 2016. https://aisel-aisnet.org/pacs2016/29 (accessed 02.2023)
23. Remane, G., Hildebrandt, B., Hanelt, A., Kolbe, L.M.: Discovering New Digital Business Model Types: A Study of Technology Start Ups from the Mobility Sector. In: PACIS 2016. https://aisel-aisnet.org/pacs2016/29 (accessed 01.2023)
24. https://commission.europa.eu/strategy-and-policy/priorities-2019-2024/europe-fit-digital-age/digital-markets-act-ensuring-fair-and-open-digital-markets_en (accessed 01.2025)
25. https://digital-strategy.ec.europa.eu/en/policies/digital-services-act-package (accessed 01.2025)
26. https://www.bmwi.de/Redaktion/DE/Publikationen/Digitale-Welt/das-project-gaia-x.html (accessed 03.2024)
27. European Union Network and Information Security Directive (NIS2). https://eur-lex.europa.eu/legal-content/EN/TXT/PDF/?uri=CELEX:02022L2555-20221227&qid=1722351127787 (accessed 03.2024)
28. National Institute of Standards and Technology Cybersecurity Framework 2.0. https://www.nist.gov/cyberframework (accessed 05.2024)
29. Alavizadeh, H., Jang-Jaccard, J., Enoch, S. Y., Al-Sahaf, H., Welch, I., Camtepe, S. A., Kim. D. D. : A Survey on Cyber Situation Awareness Systems: Framework, Techniques, and Insights. In : ACM Comput. Surv., 2022. https://www.researchgate.net/profile/Simon-Enoch/publication/360129974_A_Survey_on_Cyber_Situation_Awareness_Systems_Framework_Techniques_and_Insights/links/6267ac881b747d19c2a8f805/A-Survey-on-Cyber-Situation-

Awareness-Systems-Framework-Techniques-and-Insights.pdf?origin=journalDetail&_tp=eyJwYWdlIjoiam91cm5hbERldGFpbCJ9 (accessed 03.2024)
30. Dash B, Ansari M.F.: An Effective Cybersecurity Awareness Training Model. First Defense of an Organizational Security Strategy. In: International Research Journal of Engineering and Technology (IRJET), 2022
31. D'Arcy J, Hovac A.: Deterring Internal Information Systems Misuse. In: Communications of the ACM, Vol. 50, No. 10, pp. 113-117, 2007
32. Landress A.D, Parrish J, Terrel S.: Resilience as an Outcome of SETA Programs. In: Proceedings. 23rd American Conference in Information Systems, 2017
33. Husak M, Jirsik T, Yang S.J.: SOK: Contemporary Issues and Challenges to Enable Cyber Situational Awareness for Network Security. In: Proceedings 15th Conference on Availability, Reliability and Security, Article No. 2, pp. 1-10, 2020. https://doi.org/10/1145/3407023.3407062 (accessed 11.2023)
34. Department of the Army. FM 3-38: Cyber Electromagnetic Activities. 2014
35. Brynielson J, Franke U, Varga S.: Cyber Situational Awareness Testing. In: Combatting Cybercrime and Cyberterrorism: Challenges, Trends and Priorities, pp. 209-233, Springer Nature, 2016
36. SAFECOM: Guide to Getting Started with a Cybersecurity Risk Assessment, 2023. https://www.cisa.gov/sites/default/files/2023-02/22_safecom_guide_to_cybersecurity_risk_assessment_508-r1.pdf (accessed 02.2024)
37. SENTINELONE: What is Cybersecurity Risk Management. 2023. https://www.sentinelone.com/cybersecurity-101/cybersecurity/what-is-cyber-risk-management/ (accessed 11.2025)
38. Ross, R., Pillitteri, V., Dempsey, K., Riddle, M., Guissanie, G.: Protecting Controlled Unclassified Information in Nonfederal Systems and Organizations. NIST SP 800-171A, Rev. 3, 2024. https://csrc.nist.gov/pubs/sp/800/171/a/r3/final (accessed 03.2024)
39. The essential guide to Cybersecurity Risk Management. Fieldeffect.com, 2024. https://get.fieldeffect.com/hubfs/Resources/White%20Papers/FE-Guide-Risk%20Management.pdf (accessed 03.2024)
40. National Security Strategy – 2022. https://www.whitehouse.gov/wp-content/uploads/2022/10/Biden-Harris-Administrations-National-Security-Strategy-10.2022.pdf (accessed 03.2024)
41. The NIS2 Directive. https://www.europarl.europa.eu/RegData/etudes/BRIE/2021/689333/EPRS_BRI(2021)689333_EN.pdf (accessed 03.2024)
42. ENISA NIS Directive. https://www.enisa.europa.eu/topics/cybersecurity-policy/nis-directive-new (accessed 03.2024)
43. Luber S, Schmitz P.: Security Insider – Definition Disaster Recovery. Vogel Communication Group, 2020. https://www.security-insider.de/was-ist-disaster-recovery-a732206/ (accessed 04.2024)
44. Litone M.: Mission-Critical Network Planning. Artech House Publ. 2023
45. Marget A.: RPO and RTO: What are they and how to calculate them. Unitrends Whitepaper, 2022. https://www.unitrends.com/blog/rpo-rto (accessed 10.2023)
46. Glossary RPO. https://www.druva.com/glossary/wht-is-a-recovery-point-objective-definition-and-related-faqs/ (accessed 03.2024)
47. Salamanca F, Jimenez J. Implementing automated replication for cost effective disaster recovery. 2011. https://dsimg.ubmus.net/envelope/157842/313522/1332863421_3_21_Implementing_automated_replication_for_cost_effective_disaster_recovery (accessed 05.2023)
48. Glossary RTO. https://www.f5.com/services/resources/glossary/recovery-time-objective-rto (accessed 03.2024)
49. Kerner, S.M.: Recovery Time Objective. Techtarget Notes, 2022. https://www.techtarget.com/whatis/definition/recovery-time-objective-RTO (accessed 03.2024)
50. Testing Disaster Recovery with Chaos Engineering, 2023. https://www.gremlin.com/community/testing-disaster-recovery-with-chaos-engineering/ (accessed 03.2024)
51. Kirvan P, Sliwa C.: What is Business Impact Analysis? Techtarget Whitepaper, https://www.techtarget.com/wearchstorage/definition/business-impact-analysis (accessed 02.2024)

52. Business Impact Analysis. IT Governance Green Paper, 2022
53. Dukin J. Stellwag D.: Cost Factors of a Ransomware Attack – A Description of the Possible Threat of a Ransomware Attack on IT and OT Systems with reference to the Possible Costs with consideration of Reasonable RTO and RPO. Student Class Project, Clausthal University of Technology, Germany, 2022
54. Becker J, Knackstedt D, Pöppelbuß J.: Developing Maturity Models for IT Management – A Procedure Model and its Application. In: Business Information Systems Engineering, pp. 213-222, 2009
55. Jeston J, Nelis C.: Business Process Management: Proactical Guidelines to Successful Implementation. Elsevier Publ., 2014
56. Venkatraman V.: The Digital Matrix: New Rule for Business Transformation through Technology. Life Tree Book Publ. 2017
57. Rogers D.I.: The Digital Transformation Playbook: Rethink your Business for the Digital Age. Columbia University Press, 2006
58. National Institute of Standards and Technology (NIST) Cybersecurity Framework 2.0, 2023. https://www.nist.gov/cyberframework (accessed 10.2023)
59. CISA Secure by Design Pledge. https://www.cisa.gov/resources-tools/resources/cisa-secure-design-pledge (accessed 02.2025)
60. CISA Secure by Design Pledge. https://www.cisa.gov/sites/default/files/2024-05/CISA%20Secure%20by%20Design%20Pledge_508c.pdf (accessed 02.2025)
61. Mukkamata, S.: Security by Default: The Crucial Complement to Secure by Design. Ivanti, 2024. https://www.ivanti.com/blog/security-by-default-the-crucial-complement-to-secure-by-design (accessed 02.2025)
62. What Is Secure Software Development Lifecycle (Secure SDLC)? https://www.paloaltonetworks.com/cyberpedia/what-is-secure-software-development-lifecycle (accessed 02.2025)
63. Stouffer, K., Pillitteri, V., Pease, M., Lightman, S., Zimmernan, T., Saravia, A., Sherule, A., Thompson, M.: The Guide to Operational Technology (OT) Security. NIST SP 800-82r3, 2023. https://doi.org/10.6028/NIST.SP.800-82r3 (accessed 01.2024)
64. Peerenboom, J. P.: Infrastructure Interdependencies: Overview of Concepts and Terminology. NSF/OSTP Workshop on Critical Infrastructure: Needs in Interdisciplinary Research and Graduate Training, 2001
65. Rinaldi, S. M., Peerenboom, J. P., Kelly, T. K.: Identifying, Understanding, and Analyzing Critical Infrastructure Interdependencies. In: IEEE Control Systems Magazine, Vol. 21, No. 6, pp. 11-25, 2001
66. https://www.fortinet.com/resources/cyberglpssary/operational-security (accessed 03.2024)
67. https://securityscorecard.com/blog/how-to-write-third-party-risk-management-policies-term-and-procedures (accessed 06.2024)
68. Parra, D.: Unraveling the Connection: CIA Triad Principles and the NIST Incident Response Lifecycle in Cybersecurity. In: Medium Staff, 2023. https://medium.com/@diparra801/unraveling-the-connection-cia-triad-principles-and-the-nist-incident-response-lifecycle-in-63acde31ae14 (accessed 10.2023)
69. Johnson, C., Badger, L., Waltermire, D., Snyder, J., Skorupka, C.: Guide to Cyber Threat Information Sharing. NIST Special Publication 800-150, 2016. https://nvlpubs.nist.gov/nistpubs/SpecialPublications/NIST.SP.800-150.pdf
70. Olsen, R.: Understanding Cyber Resilience: Key Principles of a Modern Security Strategy. 2024. https://angle.ankura.com/post/102izz9/understanding-cyber-resilience-key-principles-of-a-modern-security-strategy (accessed 11.2024)
71. Federal Cybersecurity Research and Development Strategic Plan. Executive Office of the President of the United States, 2019
72. Federal Cybersecurity Research and Development Strategic Plan. Executive Office of the President of the United States, 2024
73. The NIST Cybersecurity Framework (CSF) 2.0, 2024. https://nvlpubs.nist.gov/nistpubs/CSWP/NIST.CSWP.29.pdf (accessed 05.2024)

References

74. Gartner Artificial Intelligence, 2024. https://www.gartner.com/en/information-technology/glossary/adaptive-ai
75. Gartner Artificial Intelligence, 2024. https://webinar.gartner.com/689201/agenda/session/1539147?login=ML (accessed 01.2024)
76. Gartner Artificial Intelligence, 2024. https://www.gartner.com/en/information-technology/glossary/composite-ai (accessed 11.2024)
77. Gartner Artificial Intelligence, 2024. https://www.gartner.com/en/topics/generative-ai (accessed 11.2024)
78. Kurtz, G.: 2025 Global Threat Report. CrowdStrike 2025
79. Möller, D. P. F., Haas, R. E.: Guide to Automotive Connectivity and Cybersecurity – Trends, Technologies, Innovations and Applications. Springer Nature 2019
80. Möller, D. P. F.: Machine Learning and Deep Learning for Cybersecurity in Smart Environments. In: Empowering AI Applications in Smart Life and Environments, Chapter 6, by Khalifa, N. E. M., and Taha, M, H, A. (Eds.), Springer Nature 2025
81. Machine Learning in Cybersecurity. https://www.sailpoint.com/identity-library/how-ai-and-machine-learning-are-improving-cybersecurity (accessed 01.2024)
82. GenAI planning Workbook: 4 steps to implementing generative AI in your enterprise. Gartner, 2023
83. Kutsal, B.: Cybercrime in AI services is increasing - LLMjacking attacks hit new AI model DeepSeek. 2025. https://www.bigdata-insider.de/llmjacking-angriffe-treffen-neues-ki-modell-deepseek-a-37011c7d44b065c3d5f3ef9e1e3ca131/?cmp=nl-e1f43822-bd1a-4d77-b778-d0edb893992f&uuid=8EF7B979-9559-4410-8E82-69834DD08487 (accessed 02.2025)

Chapter 2
Network and Information Security (NIS2)

2.1 Introduction to Network and Information Security (NIS2)

In today's rapidly evolving digital landscape with an escalating array of sophisticated cyber risks, cybersecurity has become paramount for organizations and societies across the globe. To be better prepared for unprecedented cyber threat risk incidents, cybersecurity measures are essential to achieve cyber resilience to Industrial Control Systems (ICS), Operational Technology (OT), Business Continuity (BC), and others. Therefore, national governments and standardization bodies developed cybersecurity frameworks and directives such as the US governmental National Institute of Standards and Technology Cybersecurity Framework (NIST CSF), the International Standard Organization (ISO) and International Electrotechnical Commission (IEC) ISO/IEC 27001 framework, or the European Union (EU) Network and Information Security Directive (NIS), to name a few.

NIST CSF 2.0 is a set of cybersecurity best practices and recommendations that makes it easier to understand cyber risks and serve as a roadmap to effectively manage and mitigate cyber risks. The competencies of NIST CSF 2.0 are aligned with six core functions that provides a structured approach to governing, identifying, protecting, detecting, responding and recovering from cyber threat risk incidents, ever-increasingly essential to organizations in any sector or community, to be combatted proactively and effectively. The path to full compliance with NIST CSF 2.0 requires a nuanced understanding to analyze how the framework apply to an organization's unique digital technology landscape.

ISO 27001, also known as ISO/IEC 27001:2022, is an information security standard created by the International Organization for Standardization (ISO) and International Electrotechnical Committee (IEC), which provides a framework and guidelines for establishing, implementing and managing an Information Security Management System (ISMS). With it, it's a global standard for secure and effective information management that defines the requirements an ISMS must meet, and proactively continuous improving it.

The EU Network and Information Security Directive, known as NIS, was a legislative framework designed to enhance the cybersecurity of network and information systems within the European Union (EU), to achieve a high common level of cybersecurity across the EU-Member States. The scope of NIS focus on targeted operators of essential (sometimes also named high critical) sectors such as: energy, transport, banking, financial market infrastructures, healthcare, and wastewater management. Hence, the background of NIS was the growing demand in cybersecurity in the era of digitalization and digital transformation with its manifold of digital-technologies based sectors and infrastructure resources. Given that fact, the NIS compliance outlined basic cybersecurity measures. However, incident reporting requirements were vague, e.g. any incident that substantially impacts the provision of services. Furthermore, the enforcement was left largely to individual EU-Member States, which leads to inconsistencies. With it, NIS require further development that end up in NIS2 [1–4], an enhanced EU cybersecurity directive with the focus on Critical Entities Resilience (CER), a network of Security Operations Centers (SOCs) and proactive and effective cybersecurity measures. Another reason was that the cyber risk landscape is never static and changed considerably and in an ever-sophisticated way. In this sense, the Network and Information Security Directive NIS2 aims to enforce a higher level of cybersecurity for Western Europe EU-Member States with broader scopes, stricter requirements, sterner sanctions and others, ensuring to remain cyber resilient in the face of evolving cybersecurity challenges. In this regard, cyber resilience is the ability of an organization to enable Business Continuity (BC) by preparing for, responding to, and recovering swiftly from cyber threat risk incidents. Hence, a cyber resilient organization can adapt to known and unknown cyber risks, and others. With NIS2, network and information security still cover the original sectors from NIS but adds entirely new categories:

- **Sectors of High Criticality**: Energy, Transport, Banking, Financial Market Infrastructures, Health, Drinking Water, Waste Water, Digital Infrastructure, ICT Service Management, Public Administration, and Space.
- **Other Critical Sectors:** Postal and Courier Services, Waste Management, Manufacture, Production and Distribution of Chemicals, Production, Processing and Distribution of Food, Manufacturing, Digital Providers, Research.

High critical and critical sectors require various measures for control that refer to:

- Increased enforcements and penalties (Article 36) for a high common cybersecurity level across the EU-Member States, based on stricter supervisory measures, including the power for national authorities to impose and conduct audits to ensure compliance. Cyber compliance enables ensuring that an organization adheres to regulations, standards, and laws related to cybersecurity and data privacy. The penalties provided for shall be effective, proportionate and dissuasive,
- Supervision for medium and large organizations in the EU expanded in its list of key sectors covered from 8 (NIS) to 18 (NIS2),

2.1 Introduction to Network and Information Security (NIS2)

- Establishment of a higher level of mandatory, reviewable and sanctionable cybersecurity measures include 23 rules for cybersecurity risk-management, cybersecurity governance, cybersecurity risk incident reporting and recovery, cyber resilience, network and information system security, and application security,
- Risk assessments and reviews of cybersecurity practices for major connected third-party service providers.

Given that fact, NIS2 sets out in total 46 Articles for its regulatory directive and lays down cooperation mechanisms among relevant authorities in each EU-Member State. It expands the scope of sectors and activities subject to cybersecurity measures, and improves their enforcement. Moreover, it applies to public administration at central and regional level and sets stricter cybersecurity obligations for EU-Member States when it comes to supervision. It also improves the enforcement of those obligations, including by harmonizing sanctions/penalties across EU-Member States. Thus, NIS2 aims to improve cooperation between EU-Member States under the umbrella of the EU Agency for Cybersecurity (ENISA) to increase cyber resilience in the EU-Member States. Form a general perspective. ENISA is working with EU-Member States to identify best EU-wide practices in their efforts to implement the NIS2 Directive. Moreover, ENISA works with EU-Member State organizations and businesses to strengthen trust in the digital transformed economy, boost the resilience of EU infrastructure resources, and ultimately keep EU citizens digitally safe, by means of sharing knowledge, developing staff and structures, and raising cybersecurity situational awareness.

Essential measures for cyber resilience in EU organizations are [5]:

- **Identity Governance and Administration:** Major concern of a cyber resilience strategy is to ensure that only the right people get the right access to the right resources at the right time, to reduce cyber risks of suffering cyber threat risk incidents. This requires to grant access rights by defined roles, rules and cybersecurity policies.
- **Privileged Access Management:** Cybercriminal activists target privileged accounts to provide access to sensitive digital data and assets and/or control over vital systems of the targeted organization. To thwart malicious cyber threat risk incidents, govern and monitor privileged access is essential, which requires capabilities such as granular delegation of administrative access, password vault to cybersecure privileged credentials, and cybersecurity threat risk incident analytics by continuous monitoring and recordings of privileged sessions.
- **Hybrid Active Directory Security and Management:** Assuming an Active Directory (AD) is down that also means Business Continuity (BC) is down. Therefore, it's vital be able to defend against, detect and recover from cyber threat risk incidents that affect AD. Therefore, it is essential to make sure preventing cybercriminals from manipulating critical groups or Group Policy Object (GPO) settings; identify and mitigate cyber threat risk incident paths that cybercriminals could not do to seize control of the domain; detect and respond to Indicators of Compromise (IoC), and be prepared to swiftly restore AD domain or forest. IoC represent digital forensic that suggests that an endpoint, network and/or information system has been breached.

- **Unified Endpoint Management:** Many cyber threat risk incidents begin on endpoints. Accordingly, it's vital to know exactly what endpoints used to manage, cybersecure and patch them effectively. In addition, continuous auditing and intelligent alerting on anomalous and suspicious activity is required to respond promptly to cyber threat risk incidents.
- **Backup and Disaster Recovery:** Not all cyber threat risk incidents can be prevented, so it's essential to have a comprehensive backup and recovery plan that accelerates disaster recovery across on-prem to cloud or hybrid environments. Ensure that the solution chosen contain cybersecure storage of backups out of the reach of cybercriminals, flexible recovery options, and safeguards against malicious code re-infecting network and information systems upon restoration from backup.

In general, the EU NIS2 Directive focuses on enhancing the cybersecurity posture of sectors that are crucial and critical to EU's digital transformed economy, categorized within the Directive as either high critical and critical, based on their relevance to EU's internal market and the potentially societal and economic consequences of disruptions to their services. NIS2 high critical sectors and subsectors are shown in Table 2.1 [6] and for critical sectors in Table 2.2 [6].

The sectors listed in both tables must adopt appropriate and proportionate technical and organizational measures to manage cyber risks, to safeguard network and information systems, and others. Therefore, NIS2 mandates as compliance a set of minimum cybersecurity requirements that all sectors must implement, which covers:

Table 2.1 High critical sectors with subsectors

No.	High critical sectors	Subsectors
1	Energy	(a) Electricity (b) District heating and cooling (c) Oil (d) Gas (e) Hydrogen
2	Transportation	(a) Air (b) Rail (c) Water (d) Road
3	Banking	
4	Financial Market Infrastructures	
5	Health	
6	Drinking Water	
7	Wastewater	
8	Digital Infrastructure	
9	ICT Service Management (B2B)	
10	Public Administration	
11	Space	

2.1 Introduction to Network and Information Security (NIS2)

Table 2.2 Critical sectors with subsectors

No.	Critical sectors	Subsectors
1	Postal and courier services	
2	Waste management	
3	Manufacture, production and distribution of chemicals	
4	Production, processing and distribution of food	
5	Manufacturing	(a) Manufacture of medical devices and in vitro diagnostic medical device (b) Manufacture of computer, electronic and optical products (c) Manufacture of electrical equipment (d) Manufacture of machinery and equipment n.e.c. (includes manufacture machinery and equipment that act independently on materials either mecha-nically or thermally or perform operations on materials, including their mechanical components that produce and apply force, and any specially manufactured primary parts.) (e) Manufacture of motor vehicles, trailers and semitrailers (f) Manufacture of other transport equipment
6	Digital providers	
7	Research	

- **Risk-Management:** Process of identifying cyber risks, analyzing the probability of impacts, prioritizing the cyber risk likelihood, treating and monitoring cyber threat risk incidents to organization's data, assets, capital, earnings and operations that can stem from a variety of sources.
- **Incident Response Plan:** Document that outlines roles and responsibilities to provide guidance on organization's activities required to be prepared before, during, and after a confirmed or suspected cyber threat risk incidents to respond and recover from it.
- **Supply Chain Security:** Risk-management that focuses on cyber threat risk incident management of external suppliers, vendors and others, by means of identifying, analyzing, and mitigating cyber threat risk incidents associated outside organization's as part of the supply chain.
- **Vulnerability Reporting:** Details of cybersecurity weaknesses discovered in a vulnerability assessment, e.g. penetration testing (pen test), to ensure a better state of cybersecurity preparedness. Pen test is a cybersecurity exercise where a cybersecurity expert attempts to find and exploit vulnerabilities in network and information systems. The purpose to simulate cyber threat risk incidents is to identify any weak spots in a network and information system's defenses, which cybercriminals could take advantage of.

Furthermore, NIS2 mandates more cybersecurity requirements that all sectors must implement. Thus, all EU organizations should have an established process to ensure proper reporting to authorities within a set timeframe after confirmed or suspected cyber threat risk incidents, which include:

- Send early warning within 24 h,
- Complete initial assessment within 72 h,
- Prepare a final, detailed report, not later than 1 month after the initial assessment.

In this context, an important issue of NIS2 is the responsibility to act against cyber threat risk incidents that requires effective boundary constraints for high critical and critical sectors. For high critical sector and critical sector organizations the amendments in NIS2 are illustrated in Table 2.3.

It is important to note that organizations which exceed the thresholds for high critical sectors but still do not fall under high critical sectors must comply with the directive in accordance with the requirements for high critical sectors. Some entities regardless of their size, like parts of the digital infrastructure and/or public administration, sole providers, Critical Infrastructure Systems (CRITIS), are also belong to the foregoing boundary constraints.

While NIS2 brings numerous benefits by its challenges of implementation and compliance, it also outlines distinct repercussions for non-compliance, comprising both monetary and non-monetary penalties, and possible sanctions. Basic sanctions/penalties (Article 36) for breaches of cybersecurity measures and reporting obligations are the most important ones. These penalties are for sector specific organization's that, for example, fail to meet agreed deadlines, which can take different forms, such as non-monetary remedies, fines and sanctions/penalties. The type of penalty varies depending on the organization and the discrepancy between required and actual implementation. National supervisory authorities can use non-monetary remedies, including compliance orders, binding instructions, implementing orders for cybersecurity audits and orders for cyber threat risk incident reporting to the organization's customers.

When it comes to sanctions/penalties, NIS2 distinguishes between high critical and critical sectors. Under NIS2, national authorities can impose a maximum sanction/penalty on high critical sectors and critical sectors, as illustrated in Table 2.4.

In contrast to the previous NIS, the NIS2 Directive shifts responsibility for the implementation and enforcement of cybersecurity measures away from the IT department and makes the top management responsible. Western Europe EU-Member States can now hold C-level managers liable if cyber threat risk

Table 2.3 High critical and critical sectors with size and revenue

High critical sectors	Critical sectors
l>250 employees	50–2490 employees
>50 million turnovers	10–50 million turnovers
>43 million balance sheets	10–43 million balance sheets

2.1 Introduction to Network and Information Security (NIS2)

Table 2.4 Sanctions/penalties of high critical and critical sectors

High critical sectors	Critical sectors
Sanctions/penalties	Sanctions/penalties
lmax EUR ten million or 2% global turnover	Max EUR seven million or 1,4% global turnover

Table 2.5 NIS2 chapters, articles, annexes, sectors, and headlines

Chapter	Articles	Headline
I	1–6	General provisions
II	7–13	Coordinated cybersecurity frameworks
II	14–19	Cooperation at union and international level
IV	20–25	Cybersecurity risk-management measures and reporting obligations
V	26–28	Jurisdiction and registration
VI	29–30	Information sharing
VII	31–37	Supervision and enforcement
VIII	38–39	Delegated and implementing acts
IX	40–46	Final provisions

Annex	Sectors	Headline
I	1–11	Sectors of high criticality
II	1–7	Other critical sectors
III		Correlation table

incidents occur due to gross negligence on the part of sector specific organizations, with sanctions/penalties varying. Compliance violations can be made public or the natural or legal person(s) responsible for the violation can be named. If it is a high critical entity, persons can be removed from all management positions if they repeatedly violate the law.

Given the increasing complexity and sophistication of cyber threat risk incidents, it is crucial for high critical and critical sector organization's to implement at least basic cyber hygiene practices. As basis for protecting high critical infrastructure, organization's should implement basic cybersecurity measures such as regular software and hardware updates, regular password changes, securing managing new installations, restricting account access at the administrator level and backing up data. Since sophisticated cyber threat risk incidents occur by means of unprecedented connectivity by digitalization, cybersecurity training and raising employee cybersecurity situational awareness is essential. This is one way to create a proactive cybersecure environment and the awareness needed to strengthen cybersecurity of operators of high critical sector services in the EU-Member States.

All foregoing and other cases are part of the NIS2 Directive, which includes 144 preambles, 9 chapters with 46 articles, and 3 annexes. The chapters with their articles and headlines, and the annexes are illustrated in Table 2.5.

The full document of the EU directive 2022/2555 of the European Parliament and of the Council of December 14 in 2022 on measures for a high common level of cybersecurity across the Union, amending EU Regulation No 910/2014 and EU Directive 2018/1972, and repealing EU Directive 2016/1148 (NIS2 Directive) can be downloaded completely with paragraphs, articles and annexes from the website: https://eur-lex.europa.eu/legal-content/EN/TXT/PDF/?uri=CELEX:32022L2555, which is recommended to have the full text at hand to read. In the following NIS2 Chapters in all text document the most pertinent paragraphs of the respective articles are given and in special cases additional and detailed textual supplementary depiction with practical, and application oriented suggested solutions described. From the general perspective, NIS2 Chapter IV is the most r one, and give essential hints to discuss practical, technical and application-oriented solutions and consultancy skills to cybersecurity risk-management measures and reporting obligations.

2.2 Chapter I General Provisions

NIS2 Chapter I introduce to the 6 Articles of General Provisions [6]. In this book chapter the term high critical and critical sectors in the articles used instead of essential and important, to stay consistent with the designation in Annex I and Annex II of the NIS2 Directive.

Article 1 Subject Matter Lays down in two paragraphs the following needs:

(i) measures that aim to achieve a high common level of cybersecurity across the European Union, with a view to improving the functioning of the internal market.
(ii.a) obligations that require EU-Member States to adopt national cybersecurity strategies and to designate or establish competent authorities, cyber crisis management authorities, single points of contact on cybersecurity (single points of contact) and Computer Security Incident Response Teams (CSIRTs);
(ii.b) cybersecurity risk-management measures and reporting obligations for sectors of a type referred to in Annex I or II as well as for sectors identified as critical entities under Directive (EU) 2022/2557;
(ii.c) rules and obligations on cybersecurity information sharing;
(ii.d) supervisory and enforcement obligations on Member States.

Article 2 Scope Consist of 14 paragraphs. Paragraphs 1 and 2 are only considered in this book chapter:

1. This Directive applies to public or private entities of a type referred to in Annex I or II which qualify as medium-sized enterprises under Article 2 of the Annex to Recommendation 2003/361/EC, or exceed the ceilings for medium-sized enterprises provided for in paragraph 1 of that Article, and which provide their services or carry out their activities within the Union.
Article 3(4) of the Annex to that Recommendation shall not apply for the purposes of this Directive.

2. Regardless of their size, this Directive also applies to entities of a type referred to in Annex I or II, where:
   (a) services are provided by:
      (i) providers of public electronic communications networks or of publicly available electronic communications services;
      (ii) trust service providers;
      (iii) top-level domain name registries and domain name system service providers;
   (b) the entity is the sole provider in a Member State of a service which is essential for the maintenance of critical societal or economic activities;
   (c) disruption of the service provided by the entity could have a significant impact on public safety, public security or public health;
   (d) disruption of the service provided by the entity could induce a significant systemic risk, in particular for sectors where such disruption could have a cross-border impact;
   (e) the entity is critical because of its specific importance at national or regional level for the particular sector or type of service, or for other interdependent sectors in the Member State;
   (f) the entity is a public administration entity:
      (i) of central government as defined by a Member State in accordance with national law; or.
      (ii) at regional level as defined by a Member State in accordance with national law that, following a risk-based assessment, provides services the disruption of which could have a significant impact on critical societal or economic activities.

Article 3 High Critical and Critical Entities Consist of 6 paragraphs. Paragraph 1 and its subparagraphs are only considered in this book chapter.

1. For the purposes of this Directive, the following entities shall be considered to be essential entities:
   (a) entities of a type referred to in Annex I which exceed the ceilings for medium-sized enterprises provided for in Article 2(1) of the Annex to Recommendation 2003/361/EC;
   (b) qualified trust service providers and top-level domain name registries as well as DNS service providers, regardless of their size;
   (c) providers of public electronic communications networks or of publicly available electronic communications services which qualify as medium-sized enterprises under Article 2 of the Annex to Recommendation 2003/361/EC;
   (d) public administration entities referred to in Article 2(2), point (f)(i);
   (e) any other entities of a type referred to in Annex I or II that are identified by a Member State as essential entities pursuant to Article 2(2), points (b) to (e);

(f) entities identified as critical entities under Directive (EU) 2022/2557, referred to in Article 2(3) of this Directive;
(g) if the Member State so provides, entities which that Member State identified before 16 January 2023 as operators of essential services in accordance with Directive (EU) 2016/1148 or national law.

Article 4 Sector-Specific Union Acts Consist of 3 paragraphs. Paragraphs 1 and 2 are only considered in this book chapter.

1. Where sector-specific Union legal acts require essential or important entities to adopt cybersecurity risk-management measures or to notify significant incidents and where those requirements are at least equivalent in effect to the obligations laid down in this Directive, the relevant provisions of this Directive, including the provisions on super vision and enforcement laid down in Chapter VII, shall not apply to such entities. Where sector-specific Union legal acts do not cover all entities in a specific sector falling within the scope of this Directive, the relevant provisions of this Directive shall continue to apply to the entities not covered by those sector-specific Union legal acts.
2. The requirements referred to in paragraph 1 of this Article shall be considered to be equivalent in effect to the obligations laid down in this Directive where:
   (a) cybersecurity risk-management measures are at least equivalent in effect to those laid down in Article 21(1) and (2); or
   (b) the sector-specific Union legal act provides for immediate access, where appropriate automatic and direct, to the incident notifications by the CSIRTs, the competent authorities or the single points of contact under this Directive and where requirements to notify significant incidents are at least equivalent in effect to those laid down in Article 23(1) to (6) of this Directive.

Article 5 Minimum harmonization This Directive shall not preclude Member States from adopting or maintaining provisions ensuring a higher level of cybersecurity, provided that such provisions are consistent with Member States' obligations laid down in Union law.

Article 6 Definitions For the purpose of this Directive, 41 definitions apply that are not considered in this book chapter.

2.3 Chapter II Coordinated Cybersecurity Frameworks

NIS2 Chapter II introduce to the 7 Articles of Coordinated Cybersecurity Frameworks [6].

Article 7 National Cybersecurity Strategy Consist of 4 paragraphs. Paragraph 1 and its 8 subparagraphs are only considered in this book chapter.

1. Each Member State shall adopt a national cybersecurity strategy that provides for the strategic objectives, the resources required to achieve those objectives,

2.3 Chapter II Coordinated Cybersecurity Frameworks 85

and appropriate policy and regulatory measures, with a view to achieving and maintaining a high level of cybersecurity. The national cybersecurity strategy shall include:

(a) objectives and priorities of the Member State's cybersecurity strategy covering in particular the sectors referred to in Annexes I and II;
(b) a governance framework to achieve the objectives and priorities referred to in point (a) of this paragraph, including the policies referred to in paragraph 2;
(c) a governance framework clarifying the roles and responsibilities of relevant stakeholders at national level, underpinning the cooperation and coordination at the national level between the competent authorities, the single points of contact, and the CSIRTs under this Directive, as well as coordination and cooperation between those bodies and competent authorities under sector-specific Union legal acts;
(d) a mechanism to identify relevant assets and an assessment of the risks in that Member State;
(e) an identification of the measures ensuring preparedness for, responsiveness to and recovery from incidents, including cooperation between the public and private sectors;
(f) a list of the various authorities and stakeholders involved in the implementation of the national cybersecurity strategy;
(g) a policy framework for enhanced coordination between the competent authorities under this Directive and the competent authorities under Directive (EU) 2022/2557 for the purpose of information sharing on risks, cyber threats, and incidents as well as on non-cyber risks, threats and incidents and the exercise of supervisory tasks, as appropriate;
(h) a plan, including necessary measures, to enhance the general level of cybersecurity awareness among citizens.

Article 8 Competent Authorities and Single Points of Contact Consist of 6 paragraphs. Paragraph 1 is only considered in this book chapter.

1. Each Member State shall designate or establish one or more competent authorities responsible for cybersecurity and for the supervisory tasks referred to in Chapter VII (competent authorities).

Article 9 National Cyber Crisis Management Frameworks Consist of 5 paragraphs. Paragraph 4 and its subparagraphs are only considered in this book chapter.

4. Each Member State shall adopt a national large-scale cybersecurity incident and crisis response plan where the objectives of and arrangements for the management of large-scale cybersecurity incidents and crises are set out. That plan shall lay down, in particular:

(a) the objectives of national preparedness measures and activities;
(b) the tasks and responsibilities of the cyber crisis management authorities;
(c) the cyber crisis management procedures, including their integration into the general national crisis management framework and information exchange channels;

(d) national preparedness measures, including exercises and training activities;
(e) the relevant public and private stakeholders and infrastructure involved;
(f) national procedures and arrangements between relevant national authorities and bodies to ensure the Member State's effective participation in and support of the coordinated management of large-scale cybersecurity incidents and crises at Union level.

Article 10 Computer Security Incident Response Teams (CSIRTs) Consist of 10 paragraphs. Paragraph 1 is only considered in this book chapter.

1. Each Member State shall designate or establish one or more CSIRTs. The CSIRTs may be designated or established within a competent authority. The CSIRTs shall comply with the requirements set out in Article 11(1), shall cover at least the sectors, subsectors and types of entity referred to in Annexes I and II, and shall be responsible for incident handling in accordance with a well-defined process.

Article 11 Requirements, Technical Capabilities and Tasks of CSIRTs Consist of 5 paragraphs. Paragraph 1 and its subparagraphs are only considered in this book chapter.

1. The CSIRTs shall comply with the following requirements.

    (a) the CSIRTs shall ensure a high level of availability of their communication channels by avoiding single points of failure, and shall have several means for being contacted and for contacting others at all times; they shall clearly specify the communication channels and make them known to constituency and cooperative partners;
    (b) the CSIRTs' premises and the supporting information systems shall be located at secure sites;
    (c) the CSIRTs shall be equipped with an appropriate system for managing and routing requests, in particular to facilitate effective and efficient handovers;
    (d) the CSIRTs shall ensure the confidentiality and trustworthiness of their operations;
    (e) the CSIRTs shall be adequately staffed to ensure availability of their services at all times and they shall ensure that their staff is trained appropriately;
    (f) the CSIRTs shall be equipped with redundant systems and backup working space to ensure continuity of their services.

The CSIRTs may participate in international cooperation networks.

Article 12 Coordinated Vulnerability Disclosure and a EUROPEAN VULNERABILITY DATABASE Consist of 2 paragraphs. Paragraph 1 and its subparagraphs are only considered in this book chapter.

1. Each Member State shall designate one of its CSIRTs as a coordinator for the purposes of coordinated vulnerability disclosure. The CSIRT designated as coordinator shall act as a trusted intermediary, facilitating, where necessary, the inter-

action between the natural or legal person reporting a vulnerability and the manufacturer or provider of the potentially vulnerable ICT products or ICT services, upon the request of either party. The tasks of the CSIRT designated as coordinator shall include:

(a) identifying and contacting the entities concerned;
(b) assisting the natural or legal persons reporting a vulnerability; and.
(c) negotiating disclosure timelines and managing vulnerabilities that affect multiple entities.

Member States shall ensure that natural or legal persons are able to report, anonymously where they so request, a vulnerability to the CSIRT designated as coordinator. The CSIRT designated as coordinator shall ensure that diligent follow-up action is carried out with regard to the reported vulnerability and shall ensure the anonymity of the natural or legal person reporting the vulnerability. Where a reported vulnerability could have a significant impact on entities in more than one Member State, the CSIRT designated as coordinator of each Member State concerned shall, where appropriate, cooperate with other CSIRTs designated as coordinators within the CSIRTs network.

Article 13 Cooperation at National Level Consist of 6 paragraphs. Paragraphs 1 and 3 are only considered in this book chapter.

1. Where they are separate, the competent authorities, the single point of contact and the CSIRTs of the same Member State shall cooperate with each other with regard to the fulfilment of the obligations laid down in this Directive.
2. Member States shall ensure that their CSIRTs or, where applicable, their competent authorities, receive notifications of significant incidents pursuant to Article 23, and incidents, cyberthreats and near misses pursuant to Article 30.
3. Member States shall ensure that their CSIRTs or, where applicable, their competent authorities inform their single points of contact of notifications of incidents, cyber threats and near misses submitted pursuant to this Directive.

2.4 Chapter III Cooperation at EU and International Level

NIS Chapter III introduce to the 7 Articles of Cooperation at EU and International Level [6].

Article 14 Cooperation Group Consist of 9 paragraphs. Paragraph 4 and its 19 subparagraphs are only considered in this book chapter.

4. The Cooperation Group shall have the following tasks:

    (a) to provide guidance to the competent authorities in relation to the transposition and implementation of this Directive;

(b) to provide guidance to the competent authorities in relation to the development and implementation of policies on coordinated vulnerability disclosure, as referred to in Article 7(2), point (c);
(c) to exchange best practices and information in relation to the implementation of this Directive, including in relation to cyber threats, incidents, vulnerabilities, near misses, awareness-raising initiatives, training, exercises and skills, capacity building, standards and technical specifications as well as the identification of essential and important entities pursuant to Article 2(2), points (b) to (e);
(d) to exchange advice and cooperate with the Commission on emerging cybersecurity policy initiatives and the overall consistency of sector-specific cybersecurity requirements;
(e) to exchange advice and cooperate with the Commission on draft delegated or implementing acts adopted pursuant to this Directive;
(f) to exchange best practices and information with relevant Union institutions, bodies, offices and agencies;
(g) to exchange views on the implementation of sector-specific Union legal acts that contain provisions on cybersecurity;
(h) where relevant, to discuss reports on the peer review referred to in Article 19(9) and draw up conclusions and recommendations;
(i) to carry out coordinated security risk assessments of critical supply chains in accordance with Article 22(1);
(j) to discuss cases of mutual assistance, including experiences and results from cross-border joint supervisory actions as referred to in Article 37;
(k) upon the request of one or more Member States concerned, to discuss specific requests for mutual assistance as referred to in Article 37;
(l) to provide strategic guidance to the CSIRTs network and EU-CyCLONe on specific emerging issues;
(m) to exchange views on the policy on follow-up actions following large-scale cybersecurity incidents and crises on the basis of lessons learned of the CSIRTs network and EU-CyCLONe;
(n) to contribute to cybersecurity capabilities across the Union by facilitating the exchange of national officials through a capacity building program involving staff from the competent authorities or the CSIRTs;
(o) to organize regular joint meetings with relevant private stakeholders from across the Union to discuss activities carried out by the Cooperation Group and gather input on emerging policy challenges;
(p) to discuss the work undertaken in relation to cybersecurity exercises, including the work done by ENISA;
(q) to establish the methodology and organizational aspects of the peer reviews referred to in Article 19(1), as well as to lay down the self-assessment methodology for Member States in accordance with Article 19(5), with the assistance of the Commission and ENISA, and, in cooperation with the Commission and ENISA, to develop codes of conduct underpinning the working methods of designated cybersecurity experts in accordance with Article 19(6);

2.4 Chapter III Cooperation at EU and International Level

(r) to prepare reports for the purpose of the review referred to in Article 40 on the experience gained at a strategic level and from peer reviews;
(s) to discuss and carry out on a regular basis an assessment of the state of play of cyber threats or incidents, such as ransomware.

The Cooperation Group shall submit the reports referred to in the first subparagraph, point (r), to the Commission, to the European Parliament and to the Council.

Article 15 CSIRTs Network Consist of 6 paragraphs. Paragraph 3 and its 16 subparagraphs are only considered in this book chapter.

3. The CSIRTs network shall have the following tasks:

   (a) to exchange information about the CSIRTs' capabilities;
   (b) to facilitate the sharing, transfer and exchange of technology and relevant measures, policies, tools, processes, best practices and frameworks among the CSIRTs;
   (c) to exchange relevant information about incidents, near misses, cyber threats, risks and vulnerabilities;
   (d) to exchange information with regard to cybersecurity publications and recommendations;
   (e) to ensure interoperability with regard to information-sharing specifications and protocols;
   (f) at the request of a member of the CSIRTs network potentially affected by an incident, to exchange and discuss information in relation to that incident and associated cyber threats, risks and vulnerabilities;
   (g) at the request of a member of the CSIRTs network, to discuss and, where possible, implement a coordinated response to an incident that has been identified within the jurisdiction of that Member State;
   (h) to provide Member States with assistance in addressing cross-border incidents pursuant to this Directive;
   (i) to cooperate, exchange best practices and aid to the CSIRTs designated as coordinators pursuant to Article 12(1) with regard to the management of the coordinated disclosure of vulnerabilities which could have a significant impact on entities in more than one Member State;
   (j) to discuss and identify further forms of operational cooperation, including in relation to:

      (i) categories of cyber threats and incidents;
      (ii) early warnings;
      (iii) mutual assistance;
      (iv) principles and arrangements for coordination in response to cross-border risks and incidents;
      (v) contribution to the national large-scale cybersecurity incident and crisis response plan referred to in Article 9(4) at the request of a Member State;

(k) to inform the Cooperation Group of its activities and of the further forms of operational cooperation discussed pursuant to point (j), and, where necessary, request guidance in that regard;
(l) to take stock of cybersecurity exercises, including those organized by ENISA;
(m) at the request of an individual CSIRT, to discuss the capabilities and preparedness of that CSIRT;
(n) to cooperate and exchange information with regional and Union-level Security Operations Centers (SOCs) in order to improve common situational awareness on incidents and cyber threats across the Union;
(o) where relevant, to discuss the peer-review reports referred to in Article 19(9);
(p) to provide guidelines in order to facilitate the convergence of operational practices with regard to the application of the provisions of this Article concerning operational cooperation.

Article 16 European Cyber Crisis Liaison Organization Network (EU-CyCLONe) Consist of 6 paragraphs. Paragraph 3 and its 5 subparagraphs are only considered in this book chapter.

1. EU-CyCLONe shall have the following tasks.

   (a) to increase the level of preparedness of the management of large-scale cybersecurity incidents and crises;
   (b) to develop a shared situational awareness for large-scale cybersecurity incidents and crises;
   (c) to assess the consequences and impact of relevant large-scale cybersecurity incidents and crises and propose possible mitigation measures;
   (d) to coordinate the management of large-scale cybersecurity incidents and crises and support decision-making at political level in relation to such incidents and crises;
   (e) to discuss, upon the request of a Member State concerned, national large-scale cybersecurity incident and crisis response plans referred to in Article 9(4).

Article 17 International Cooperation The Union may, where appropriate, conclude international agreements, in accordance with Article 218 TFEU, with third countries or international organizations, allowing and organizing their participation in particular activities of the Cooperation Group, the CSIRTs network and EU-CyCLONe. Such agreements shall comply with Union data protection law.

Article 18 Report on the State of Cybersecurity in the Union Consist of 3 paragraphs. Paragraph 1 with its 5 subparagraphs are only considered in this book chapter.

1. ENISA shall adopt, in cooperation with the Commission and the Cooperation Group, a biennial report on the state of cybersecurity in the Union and shall submit and present that report to the European Parliament. The report shall, inter alia, be made available in machine-readable data and include the following:

(a) a Union-level cybersecurity risk assessment, taking account of the cyber threat landscape;
(b) an assessment of the development of cybersecurity capabilities in the public and private sectors across the Union;
(c) an assessment of the general level of cybersecurity awareness and cyber hygiene among citizens and entities, including small and medium-sized enterprises;
(d) an aggregated assessment of the outcome of the peer reviews referred to in Article 19;
(e) an aggregated assessment of the level of maturity of cybersecurity capabilities and resources across the Union, including those at sector level, as well as of the extent to which the Member States' national cybersecurity strategies are aligned.

Article 19 Peer Review   Consist of 9 paragraphs. Paragraph 1 and its 7subparagraphs are only considered in this book chapter.

1. The Cooperation Group shall, by 17 January 2025 establish, with the assistance of the Commission and ENISA, and, where relevant, the CSIRTs network, the methodology and organizational aspects of peer reviews with a view to learning from shared experiences, strengthening mutual trust, achieving a high common level of cybersecurity, as well as enhancing Member States' cybersecurity capabilities and policies necessary to implement this Directive. Participation in peer reviews is voluntary. The peer reviews shall be carried out by cybersecurity experts. The cybersecurity experts shall be designated by at least two Member States, different from the Member State being reviewed.

The peer reviews shall cover at least one of the following:

(a) the level of implementation of the cybersecurity risk-management measures and reporting obligations laid down in Articles 21 and 23;
(b) the level of capabilities, including the available financial, technical and human resources, and the effectiveness of the exercise of the tasks of the competent authorities;
(c) the operational capabilities of the CSIRTs;
(d) the level of implementation of mutual assistance referred to in Article 37;
(e) the level of implementation of the cybersecurity information-sharing arrangements referred to in Article 29;
(f) specific issues of cross-border or cross-sector nature.

2.5 Chapter IV Cybersecurity Risk-Management Measures and Reporting Obligations

NIS2 Chapter IV introduce in 6 Articles in Cybersecurity Risk-Management Measures and Reporting Obligations [6]. In order to deeper understanding the measures in the 6 Articles, a supplementary depiction is introduced that give pertinently hints to practical, technical and application-oriented solution examples and consultancy skills for effective and efficient realization of cybersecurity risk-management. **Please note that the supplementary depiction in this book is for informational purposes only and should not be considered legal guidance.**

Due to the increasing frequency, sophistication, and severity of cyber threat risk incidents, all high critical and critical sector organizations must ensure cyber risk measures, receiving appropriate and sustainable attention within their cybersecurity risk-management plans.

2.5.1 Article 20 Governance

Article 20 Governance Consist of 2 paragraphs that describe the intended goals. In order to better understand the paragraphs some supplementary hints are given.

1. Member State shall ensure that the management bodies of essential and important entities approve the cybersecurity risk-management measures taken by those entities in order to comply with Article 21, oversee its implementation and can be held liable for infringements by the entities of that Article.

The application of this paragraph shall be without prejudice to national law as regards the liability rules applicable to public institutions, as well as the liability of public servants and elected or appointed officials.

2. Member States shall ensure that the members of the management bodies of essential and important entities are required to follow training, and shall encourage essential and important entities to offer similar training to their employees on a regular basis, in order that they gain sufficient knowledge and skills to enable them to identify risks and assess cybersecurity risk-management practices and their impact on the services provided by the entity.

The two paragraphs of Article 20 describe the necessity of measures in risk-management to comply with Article 21 and liability rules applicable to public organizations. Thus, Article 20 refers to management responsibilities in case of violations through stricter legal requirements and obligations. For this purpose, management must approve and monitor cybersecurity measures and have employees trained for potential cyber risks (cybersecurity awareness). Furthermore, management must keep themselves informed about relevant processes at an ongoing basis and check whether any indications of cyber risks or legal violations exist. In addition, management bodies (particularly in the case of negligent misconduct) must assure to be secured against cyber threat risk incidents by means of system recovery, emergency procedures, as well as established crisis action teams. Organizations must also be cyber secured to Business Continuity (BC), which is ensured in the event of cyber threat risk incidents through system recovery, emergency procedures and the established crisis response teams.

2.5.2 Article 21 Cybersecurity Risk-Management Measures

Article 21 consist of 5 paragraphs that describe the intended requirements in cybersecurity risk-management. In order to deeper understanding the 5 paragraphs, a supplementary depiction is introduced to give detailed essential hints to practical, technical and application-oriented solution examples and consultancy skills for effective and efficient realization of cybersecurity risk-management, to fulfill the requirements of Article 21.

1. Member States shall ensure that essential and important entities take appropriate and proportionate technical, operational and organizational measures to manage the risks posed to the security of network and information systems which those entities use for their operations or for the provision of their services, and to prevent or minimize the impact of incidents on recipients of their services and on other services.

Considering the state-of-the-art and, where applicable, relevant European and international standards, as well as the cost of implementation, the measures referred to in the first subparagraph shall ensure a level of cybersecurity of network and information systems appropriate to the cyber threat risk incidents posed. When assessing the proportionality of those measures, due account shall be taken of the degree of the entity's exposure to risks, the entity's size and the likelihood of occurrence of cyber threat risk incidents and their severity, including their societal and economic impact.

2. The measures referred to in paragraph 1 shall be based on an all-hazards approach that aims to protect network and information systems and the physical environment of those systems from incidents, and shall include at least the following:

The measures to Article 21.2, illustrated in Table 2.6, must met the requirements for cybersecurity risk-management. They are illustrated Table 2.7, considering some of the often-used international cybersecurity frameworks for comparison.

In order to meet the requirements of Article 21.2, practical and application effective solutions are required, offered among others by cybersecurity-specialized enterprises or can be developed and implemented on its own by organizations, in accordance with the requirements of NIS2. In case an organization decides for an own cybersecurity development, the steps described in Chap. 1, Sects. 1.3.2 to 1.3.4, can be carried out for cybersecurity risk-management planning and establishment, which follow NIS2. Given that fact, practical, technical and application-oriented hints to measures required are described in the following sections for Article 21.2.

Table 2.6 Cybersecurity risk-management measures considered in Article 21.2

Article 21.2	Measures
(a)	Policies on risk analysis and information system security
(b)	Incident handling
(c)	Business continuity, such as backup management and disaster recovery, and crisis management
(d)	Supply chain security, including security-related aspects concerning the relationships between each entity and its direct suppliers or service providers
(e)	Security in network and information systems acquisition, development and maintenance, including vulnerability handling and disclosure
(f)	Policies and procedures to assess the effectiveness of cybersecurity risk-management measures
(g)	Basic cyber hygiene practices and cybersecurity training
(h)	Policies and procedures regarding the use of cryptography and, where appropriate, encryption
(i)	Human resources security, access control policies and asset management
(j)	Use of multi-factor authentication or continuous authentication solutions, secured voice, video and text communications and secured emergency communication systems within the entity, where appropriate

Table 2.7 Requirements from Article 21.2

Article 21.2	Requirement
(a)	Guidelines for risks and information security: NIST CSF, ISO/IEC 27001 Annex 16
(b)	Prevention, detection and management of cyber incidents: ISO/IEC 27001 Annex 16
(c)	Business continuity management (BCM), disaster recovery, crisis management: ISO/IEC 27001
(d)	Security in the supply chain, up to cyber secure development at suppliers' site: ISO/IEC 27001
(e)	Security in the procurement of IT and network systems
(f)	Guidelines for measuring and evaluating cybersecurity and cyber risk measures
(g)	Cybersecurity awareness and training, social-engineering security
(h)	Specifications for cryptography and, if possible, encryption
(i)	Human resources security, access management, asset management, information systems management system (ISMS): ISO 27001 Annex 8
(j)	Multi-factor authentication (MFA), single sign-on (SSO), secure voice, video and text communication, secure emergency communication system(s)

Measures to Article 21.2.a: Policies on Risk Analysis and Information System Security

A policy in risk analysis and information system security is a document that provides a set of guidelines for an organization to identify vulnerabilities, and manage cyber risks that could impact its assets, data, information systems, objectives, operations, or reputation. In this sense, the policy offers a structured approach to risk analysis and management, guiding the organization's efforts and fostering a cyber risk-aware culture to stay cyber secure by achieving the CIA Triad (see Sect. 1.5). Risk analysis includes vulnerability assessments a well as penetration tests to identify and address potential cybersecurity weaknesses that must be protected in the face of today's ever sophisticated cyber threat risk incidents. In this context, a vulnerability assessment involves, for example, an active network examination to identify hosts, software, and configurations with vulnerabilities that have not been remediated so far. With this test many vulnerabilities be identified, including unpatched software, network protocols using outdated encryption and security standards, or exposed ports and network services not adequately protected behind a firewall. In contrast, penetration tests involve hiring a cybersecurity professional or an ethical hacker, attempting to compromise a network as a cyber threat risk incident cybercriminal would do. Pen testers use a combination of automated tools, manual testing, and individual skills to identify vulnerabilities, then attempt to exploit them to gain access to additional hosts, accounts, and permissions. This process is repeated to bore into a network and identify vulnerabilities hidden from surface assessments [7]. In his sense, a proactive continuous monitoring and a comprehensive cybersecurity plan is required, as part of an Information Security Management System (ISMS). An ISMS comprises a set of policies and procedures to ensure, manage, control, and continuously improve information security in organizations, with the scope to preserve the characteristics of information described in the CIA triad.

An international standard for information security and hence for creating an ISMS is ISO/IEC 27001. The ISO/IEC 27001specifies the requirements for an ISMS to ensure the CIA triad of all corporate digital data, including:

- Intellectual property,
- Financial information,
- Personally, identifiable information,
- Information managed by third parties,
- And others.

Following ISO/IEC 27001 to build a comprehensive and effective ISMS, an effective audit program is required keeping information assets safe and secure and complying with the regulatory requirements in NIS2. Thus, ISO/IEC 27001 is an ideal resource for organizations looking to bolster their cybersecurity practices and mitigate their cyber threat risk incidents.

However, to be successful to build a comprehensive and effective ISMS, the management must demonstrate leadership and commitment with respect to the ISMS. Specifically, the management must:

- Provide appropriate resources, financial-wise and staff-wise,
- Establish policies and objectives organization-wide,
- Ensure the ISMS achieves its intended outcomes,
- Ensure the ISMS is integrated into the organization-wide processes,

- Promote continual improvement, because cyber threat risk incidents improve over time,
- And others.

Another significant advantage of ISO/IEC 27001 is to achieve independent certification against it to fulfill NIS2 requirements such as:

- Documented policies and procedures,
- Continual improvement,
- Risk assessment and management.

Therefore, an ISO/IEC 27001 certification creates trust and strengthens the organization's cybersecurity image. Business partners and customers receive meaningful proof that a state-of-the-art cybersecurity maturity level has been established, and is also practiced in the organization. At the same time, the organization minimize business and liability risks and increase competitiveness in the market. An ISMS also ensures that organizations information assets, such as intellectual property, personnel or financial data and others, entrusted by customers or third parties are adequately protected against any cyber threat risk incident. Given that fact, an ISMS is some kind of contingency management that integrates information security policy, risk-management, control, continuous improvement against constantly evolving cyber threat risk incidents, and internal auditing, consistent with organizational objectives, as shown in Fig. 2.1, to stay ahead of the competition and thrive it, because the architecture in Fig. 2.1 use the individual components and unites them, creating a cohesive whole.

An ISO 27001 certification to an ISMS requires that the organization identifies the digital data and assets, and provides the following assessment:

- Risks the information assets face,
- Steps taken to protect the information assets,
- Plan of action in case a cyber treat risk incident happens,
- Identification of responsible person(s) for each step of the information security process,
- And others.

Given that fact, ISMS offers numerous benefits, such as [8]:

- **Protect Sensitive Data:** Including intellectual property data, financial data, customer data, data entrusted to companies through third parties, personal data, and other.
- **Meet Regulatory Compliance:** Enables organization to meet regulatory compliance such as laws, regulations, guidelines and specifications relevant to its business processes, contractual requirements and others. Violation of legal regulations and directives comes with sanctions/penalties, e.g., NIS2, thus an ISMS be beneficial.
- **Provide Business Continuity.** With an ISMS, organizations automatically increase their level of defense against cyber threat risk incidents. This reduces the number of cybersecurity incidents, which result in fewer disruptions and less downtime as important factors for maintaining business continuity.

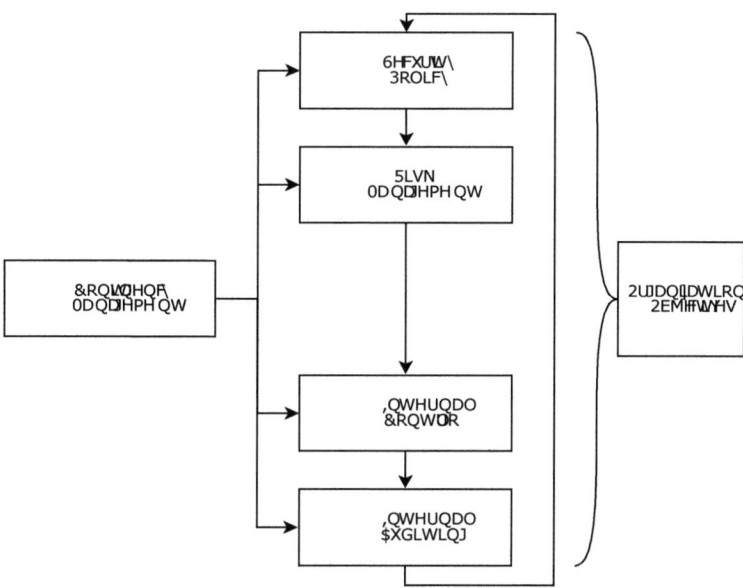

Fig. 2.1 Diagram of contingency measures, based on the contingency management block that feed at least security policy, risk management, internal control and internal auditing that finally result in the output organizational objectives

- **Reduce Costs.** ISMS enable a thorough risk assessment of all assets, which offers organizations to prioritize the highest risk assets to prevent indiscriminate spending on unneeded defenses and provide a focused approach toward cyber securing them. Thus, this structured approach significantly cuts organization's total spending, along with less downtime due to a reduction in cyber threat risk incidents.
- **Enhance Organizations Culture.** ISMS provide an efficient approach for cybersecurity and asset management throughout the organization. This encourages all employees to cybersecurity situational awareness (see Sect. 1.3.1) to understand the risks tied to information assets and adopt cybersecurity best practices as part of their daily routines.
- **Adapt to Emerging Cyber Threat Risk Incidents:** ISMS enable organizations prepare and adapt to the constantly evolving cyber threat risk incidents and the continuously changing demands of the cybersecurity landscape.

Therefore, ISMS determine the maturity of information security based on how well protection goals are achieved by the following advantages:

- Proactively address data and asset protection and cybersecurity goals,
- Achieve legal and contractual requirements like ISO/IEC 27001, NIST CSF 2.0, NIS2, and others,
- Qualify for enhanced business opportunities.

Against this background, NIST CSF 2.0 is a set of cybersecurity best practices and guidelines, developed to enable organizations to deeper dive in understanding cyber risks and improve their defenses to enhance information security. In NIST CSF 2.0 [9] the new Govern function is an overarching addition introduced to emphasize governance's importance in effectively managing cyber risks. It focuses on establishing and maintaining the governance structures necessary for an organization to manage its cyber risks in alignment with its business objectives and regulatory requirements, which is achieved by:

- Establishing risk-management objectives with agreement from organizational stakeholders,
- Establishing a standardized method for calculating, documenting, categorizing, and prioritizing cybersecurity risks,
- Establishing and maintaining risk appetite and risk tolerance statements,
- Including cybersecurity risk-management activities and outcomes in organizations risk-management,
- Establishing and communicating cybersecurity roles, responsibilities, and authorities,
- Reviewing and adjusting the cybersecurity risk-management strategy to ensure coverage of organizational risks and requirements,
- Evaluating organizational cybersecurity risk-management performance and making necessary adjustments,
- Establishing lines of communication for cybersecurity risks across the organization, including suppliers and third parties.

Measures to Article 21.2.b: Incident Handling

Any cybersecurity challenge can escalate into something bigger if not handled carefully. Without a sufficient incident handling plan for network and information systems, these technical issues can result in a problematic security breach and finally a system collapse. In this sense, incident handling is a response plan developed by organizations to counteract cyber threat risk incidents. This means it's an organized and systematized well-defined proactive reaction plan to instantly address cyber threat risk incidents is critical to enable complete recovery. Moreover, it supports organizations in cybersecurity problem finding, identification, analysis and response against of cyber threat risk incidents. Furthermore, incidents in information security are violations of the CIA triad (see Sect. 1.5) protection goals. Hence, the goal of incident handling is to mitigate and possibly, dodge the damage of a potential cyber theft within a network and information system structure. An effective incident handling can be achieved by developing and instituting effective processes and procedures for the six phases of incident response: *preparation, identification, containment, eradication, recovery, and follow-up*, whereby the incident handling process should be consistent and compatible with any forensic services that the organization may require to ensure that critical evidence is handled properly.

Also, an incident handling plan offers organizations the possibility to protect their reputation and reduce, for example, ransom costs of a possible security breach. It also allows them to have a much safer environment away from cyber threat risk incidents. As reported in [10] incident-handling capability should be available 24 h per day, 7 days a week. A service provider may be able to provide one or a combination of services.

Incident-handling services may include the following:

- Developing the incident handling program,
- Developing and maintaining system configuration profiles,
- Providing forensics capabilities,
- Testing and updating incident handling procedures,
- Managing and executing the incident handling procedures.

Incident-handling procedural services may include the following:

- Isolation of affected systems and platforms,
- Triage,
- Report assessment (interpret log files, prioritize, and analyze),
- Identification/Verification (determine nature and scope of incident),
- Categorization (determine sensitivity of compromised information, system, and incident),
- Internal and external coordination (notify and coordinate with appropriate internal and external parties on a need-to-know basis),
- Resolution,
- Technical assistance (provide detailed analysis of event, such as how it occurred and what failed to work properly),
- Eradication (eliminate incident causes and effects),
- Recovery (return systems to normal operations),
- Preventative support,
- Operations, maintenance, and monitoring of Intrusion Detection System (IDS)—incident response tied directly to IDS,
- Training,
- Post-incident consulting,
- Third-party analysis, validation of eradication and recovery, corrective and mitigation actions, and reporting.

However, it should be noted that when outsourcing IT services or products applied, the organization bears joint responsibility for this service in accordance with NIS2. Thus, the organization also must follow the three-step reporting procedure in case of cyber threat risk incidents (see Sect. 2.5.4 Article 23 Reporting obligations):

- They must give the relevant authority a first early warning within 24 h of the discovery of a cyber threat risk incident.
- The affected organization must then provide a more formal cyber threat risk incident notification within 72 h.

Table 2.8 Incident handling requirements to fulfill Article 21.2.b

Activity	Actions Taken
Preparation	Preparing for potential incidents: Define clear communication channels, implement response checklists, and qualify staff with quality cybersecurity training
Identification	Identifying and assessing incidents: Assess whether the incident is a cyber-threat or a cyber-attack, evaluate intensity, and classify the incident based on its type
Containment	Containing the impact of incidents: Isolate the affected networks and information systems and impede the incident from propagating further
Eradication	Investigating and eradicating incidents: Make sure that the incident in no longer present in the organization network and information systems by investigating the root cause of the incident and eradicating them from the organizations network and information system
Recovery	Recovering and restoring operations: Restore the affected networks and information systems to their pre-incident state to get the organization back up and running as normal
Lessons learned	Learning from the incident: Document everything that occurred during the incident and the response. Use this information to recognize areas for improvement in the organization's security and incident response plan
Ongoing improvement	Ongoing testing and evaluation: Strengthen security posture by continuously testing and evaluating the incident response plan to ensure it remains current and effective due to ever-evolving cyber risks

- A final report must be given a month after the formal cyber threat risk incident notification, and the organization must respond to any requests for status updates and provide progress reports if asked to do so.

A best practice incident handling example, based on a sequence of essential activities to be more resilient, is shown in Table 2.8 [11].

In summary, the primary purposes of an incident handling are to:

- Identify and assess incidents swiftly by conduct a thorough incident investigation to determine the root causes and implement corrective actions,
- Establish protocols for effective incident response,
- Develop and enhance the incident management processes through lessons learned and continuous improvement to improve future incident handling.

Measures to Article 21.2.c: Business Continuity, Such as Backup Management and Disaster Recovery, and Crisis Management

Business Continuity Management (BCM) integrates the disciplines Backup Management (BM), which ensure that less or no data are lost during a disaster through data backed up in a different secure storage facility, Disaster Recovery (DR) to enable technological facility continuity, and Crises Management (CM) supporting organizational/operational continuity.

BM and DR involve periodically updating copies of files, storing them in one or more remote locations, and using the copies to continue or resume business operations in the event of an outage like data loss due to file damage, data corruption, cyber threat risk incidents and others. In this context, **BM** is the process of making the file copies, while DR consists of IT technologies and best practices to prevent data loss and business disruption by using copies of files to swiftly reestablish access to applications, data and IT technology resources after an outage [12].

A DR-plan is a formal policy document created to assist an organization in executing recovery processes in response to a cybersecurity disaster to protect business information system resources against cyber threat risk incidents. The purpose of the DR-plan is to comprehensively describe the consistent actions that must be taken before, during, and after a cybersecurity disaster, so that the organizations team can take those actions. Therefore, it contains detailed instruction on strategizing, planning, deploying appropriate technology, and continuous testing, on how to respond to unexpected cyber threat risk incidents and any other disruptive events or more generally promote recovery. However, a DR-plan alone does not constitute a full disaster recovery plan. Thus, the DR also involves ensuring that adequate storage and compute is available to maintain robust failover and failback procedures. *Failover* is the process of offloading workloads to backup systems so that the organizations processes and end-user experiences are disrupted as little as possible. *Failback* involves switching back to the original primary systems. Another backup management possibility is applying Disaster Recovery-as-a-Service (DRaaS). Here, Recovery Time Objective (RTO) and Recovery Point Objective (RPO) documented in a Service-Level-Agreement (SLA) of a provider that outlines organizations downtime limits and application recovery [13]. Considering all these facts, a methodological approach that sub-divides DR into sequential steps such as disaster prevention, disaster preparedness, disaster response, and disaster recovery is effective, to bounce back after cyber threat risk incidents. However, another effective way developing a DR-plan can be done in conjunction with a Business Continuity Plan (BCP). A BCP is a complete organizational plan that consists of the components:

- **Business Resumption Plan:** Documentation of a predetermined set of instructions or procedures that describe how business processes will be recovered, resume and restored after a disruption has occurred.
- **Continuity of Operations Plan:** effort of an organizations to ensure they can continue to perform their essential functions during cyber threat risk incidents.
- **Incident Handling Plan:** Outlines actionable steps required to prepare for, respond to, and recover from a cyber threat risk incident.
- **Disaster Recovery Plan:** Document that contains detailed instructions on how to respond to unplanned cyber threat risk incidents.

Thus, the BCP describes all steps taken during a cyber threat risk incident, and outlines preventative measures that mitigate cyber threat risks. Hence, a BCP answer these questions:

- What is the plan objective, or why does the organization need it?
- What constitutes a disaster that would activate the plan?
- Who does what during a disaster?
- How will personnel communicate, and who contacts whom?
- What is the likelihood of events, such as cyber threat risk incidents, and human errors?

- What is the business impact of each of those events?
- What technologies are the organization leveraging to ensure continuity?
- What weaknesses and gaps the organization need to correct and fill?

When a BCP fit well, executives, stakeholders, and personnel know what to do and how to do it, and can easily access the plan and follow the steps as written if any confusion occurs. However, the BCP is worse when not having a recovery team to manage it, which plays the important roles in planning and carrying out organization's emergency procedures. Their responsibilities include:

- Introduce mandatory and updating the BCP,
- Identifying new cyber threat risk incident and preventative solutions,
- Training personnel mandatory on disaster response actions,
- Coordinating interdepartmental communication,
- Activating mandatory the BCP when a situation warrants it.

Furthermore, the BCP must assess the organizations unique cyber treat risk incidents, because this allows to identify the organizations vulnerabilities and to analyze the business impact, which explores how the potential disasters identified will affect the business. This allows to plan strategically and prioritize resources appropriate. Thus, a BCP categorize the impact of each cyber threat risk incident on a scale of 1 to 5, which makes it easier to gauge the severity from a high-level standpoint, particularly when comparing it against the likelihood.

In addition, the BCP includes disaster response procedures, technology, backup locations and physical assets, lines of communication, testing and mock recovery as well as periodic review and recommendation, and finally BCP update.

However, dealing with BCP, some essential terms must deeply be understood like:

- **Recovery Time Objective (RTO):** Amount of time that it takes to recover normal business operations after an outage. For RTO it is essential to consider how much time are allowed to lose data, as well as the impact this will have on BC. RTO might vary greatly from one type of business to another, from seconds, to hours, to days with regard to the acceptability or unacceptability of associated losses in revenues, and/or reputation.
- **Recovery Point Objective (RPO):** Amount of data the organization can afford to lose in a disaster. The organization might copy data to a remote data center continuously so that an outage will not result in any data loss, or might decide that another external backup storage device/location that is continuously updated fit, or losing 5 min or 1 h of data would be an acceptable risk.

Measures to Article 21.2.d: Supply Chain Security, Including Security-Related Aspects Concerning the Relationships Between Each Entity and Its Direct Suppliers or Service Providers

Supply chains make a significant contribution to maintain the global economy. Therefore, supply chains security has a positive impact on the global economy. A supply chain attack is a type of cyber threat risk incident carried out against an organization's suppliers as a means to gain unauthorized access to that organization's assets, systems or data. For example, a cybercriminal infected the Codecov Bash uploader, part of a code coverage testing tool that automatically sends reports to customers. By injecting malicious code into the script, cybercriminals eavesdropped on Codecov servers to manipulate and stole customer data. Given this fact, supply chain cybersecurity is a must have.

Supply chain cybersecurity risks can be divided from the economic perspective into cyber risk types according to their cyber risk sources and their results such as deliver cyber risks, lack of information sharing, operational risks, financial risks, and demand risks. However, due to digitalization, supply chains are affected by numerous cyber risks and disruptions. Therefore, measures to ensure supply chain cybersecurity in the form of supply chain risk-management, resilience and cybersecurity are essential. Traditional IT security, with the goal protecting the organization from cyber risks, does not offer sufficient approaches to assess and contain cyber risks arising from supply chain cyber threat risk incidents. Thus, a comprehensive supply chain risk-management strategy is crucial for ensuring organization's resilience and responsiveness to minimize supply chain cyber risks. This involves considering all cyber threat risk incidents risks that may be present throughout the supply chain, including logistics, suppliers, and locations. Properly implemented, the Supply Chain Risk-Management (SCRM) offers a significant competitive advantage. Therefore, the SCRM includes four basic stages that enable to identify, understand, and address the supply chain cyber risks to an organization, needed to prevent cyber threat risk incidents, which can be achieved through the following steps [14]:

- **Identification:** List of all potentially cyber threat risk incidents that could impact the supply chain, and identifying their location.
- **Quantification:** Evaluation of the probable impact of each cyber threat risk incident on organization's operations, finances, and reputation.
- **Mitigation:** Establishing an appropriate strategy for addressing the probale impact of each type of supply chain disruption.
- **Response:** Evaluate how fast the organizations cyber defense teams can react to disruptions and how swift the organization can recover.

A SCRM proactive model include the following two steps [15]:

- **Monitor Risk throughout the Supply Chain:** Prioritize and focus on third-parties or systems with the highest cyber risk and highest impact on organizations business. After determining high cyber risk partners or systems, the following cybersecurity controls be established:
  - **Cybersecurity updates**: Define how often the organization, or the partner organization, must update shared systems.
  - **Network cybersecurity:** Defines how to manage firewalls, cloud configurations, databases, and other critical assets on the local network or the supplier's network.
  - **Endpoint cybersecurity:** Defines how secure laptops, desktops, servers, and mobiles devices should be to access the network.
  - **Web application cybersecurity:** Defines how to protect against web-based cyber risks like cross-site scripting (XSS), SQL injection, or XML external entities (XXE).

- **Create a Risk-Aware Culture for Managing Unknown Cyberrisks.** Include employee training, to ensure that everyone in the organization understands the impact of supply chain cyber risks, and clear communication, to ensure that everyone is aware of the organization's cyber risk tolerance across the supply chain, so stakeholders can make informed decisions when working with suppliers and customers and respond swiftly to new cyber risks and mitigate them, changing processes and partner relationships if needed.

  Risk-awareness should also be part of the supplier due diligence process. By partnering with suppliers with a cyber risk-aware culture, the organization can create a cyber risk-aware supply chain that helps reduce unknown cyber threat risk incidents.

Existing cybersecurity frameworks like the global-accessible knowledge base of adversary tactics and techniques based on real world observations (MITRE ATT&CK) [16], NIST-SP 800-security and privacy controls for information systems and organizations [17] and others also address supply chain security. NIST CSF 2.0 [9] refers to cyber risks of globally distributed, extensive, and interconnected supply chains. Given their complex and interconnected relationships, SCRM is critical for organizations. Cybersecurity SCRM (C-SCRM) is a systematic process for managing exposure to cyber risks throughout supply chains and developing appropriate response strategies, policies, processes, and procedures. The Subcategories within the CSF C-SCRM Category (GV.SC) provide a connection between outcomes that focus purely on cybersecurity and those that focus on C-SCRM. SP 800-161 C SCRM Practices for Systems and Organizations, provides in-depth information on C-SCRM [18].

AS reported in [19], supply chain cybersecurity refers to the identification and management of cyber risks, through a coordinated approach among supply chain members and their third-parties, to reduce vulnerability and compromising as a whole, e.g., avoiding product manipulation or product delivery prior to receipt by a final customer for the purpose of data or system compromise. Supply chain compromise can take place at any stage of the supply chain including [20]:

- Manipulation of development tools,
- Manipulation of source code repositories,
- Manipulation of source code in open-source dependencies,
- Manipulation of software (SW) update/distribution mechanisms,
- Replacement of legitimate software (SW) with modified versions,
- Sales of modified/counterfeit products to legitimate distributors,
- And others.

While supply chain compromise can impact any component of hardware or software, cybercriminals looking to gain execution often focused on malicious additions to legitimate software in organizations software distribution or update channels [21–23]. Targeting may also be specific to a desired organization or to specific organizations [21, 24, 25]. Hence, MITRE organization recommends for supply chain security techniques the following mitigation and detection measures [15, 20], shown in Tables 2.9 and 2.10:

Table 2.9 Mitigation recommendations from Article 21.2.d

ID	Mitigation	Description
M1013	Application Developer Guidance	Application developers should be cautious when selecting third-party libraries to integrate into their application. Additionally, where possible, developers should lock SW dependencies to specific versions rather than pulling the latest version on build [25]
M1046	Boot Integrity	Use secure methods to boot a system and verify integrity of operating system and loading mechanisms
M1933	Limit SW Installation	Where possible, consider requiring developers to pull from internal repositories containing verified and approved packages rather than from external ones [26]
M1051	Update SW	A patch management process should be implemented to check unused dependencies, unmaintained and/or previously vulnerable dependencies, unnecessary features, components, files, and documentation
M1016	Vulnerability scanning	Continuous monitoring of vulnerability sources and the use of automatic and manual code review tools should also be implemented as well [27]

Table 2.10 Detection recommendations from Article 21.2.d

ID	Data source	Data component	Detects
DS0022	File	File Metadata	Use verification of distributed binaries through hash checking or other integrity checking mechanisms. Scan downloads for malicious signatures and attempt to test SW and updates prior to deployment while taking note of potential suspicious activity
DS0013	Sensor Health	Host Status	Perform physical inspection of HW for potential tampering. Perform integrity checking on pre-OS boot mechanisms for malicious purposes and compare against known good baseline behavior

Measures to Article 21.2.e: Security in Network and Information Systems Acquisition, Development and Maintenance, Including Vulnerability Handling and Disclosure

Cybersecurity measures within the lifecycle of network and information systems involves identifying and mitigating vulnerabilities during acquisition, but also implementing cybersecure practices during development; and the timely application of patches is crucial to addressing cyber risks. This requires, that organizations must be transparent about vulnerabilities and their handling with a thorough vulnerability disclosure. Therefore, a strong vulnerability management policy is needed to provide guidance for the timely identification and remediation of vulnerabilities within an organizations networks and information systems. Usually, it's impossible to eliminate all vulnerabilities; actively identifying and mitigating them reduces the chances of network and information systems compromise and limits malicious activities if a cyber threat risk incident breach occurs. From a general perspective a differentiation between passive vulnerability tracking and active vulnerability testing make sense, as described in [7]:

- **Passive Vulnerability Tracking:** Vendors often release cybersecurity updates to address known vulnerabilities. This can happen randomly, as vulnerabilities are found and fixed, or on a formal schedule. For example, Microsoft provides updates for Windows and related software on the second Tuesday of every month. Unfortunately, cybercriminals also aware on these releases. They'll act speedily to exploit software not yet patched following an update. For this reason, any known vulnerable software present on a network must be rapidly identified and updated.
- **Active Vulnerability Testing:** Passive vulnerability tracking and software updates are an extremely important issue, because unique combinations of software, systems, and services create unprecedented vulnerabilities. These cybersecurity vulnerabilities can be difficult to find without directly interacting with systems and assessing how they respond during testing. Active testing is typically referred to as penetration testing, though this term is broad and may be used to refer to several diverse types of testing activities.

  Penetration tests involve cybersecurity professionals or ethical hackers, attempting to compromise a network as a cybercriminal would. Pen testers use a combination of automated tools, manual testing, and individual skills to identify vulnerabilities, then attempt to exploit them to gain access to additional hosts, accounts, and permissions. This process is repeated to bore into a network and identify vulnerabilities hidden from surface assessments [7].
- **Vulnerability Disclosure:** Process of reporting cybersecurity vulnerabilities in networks, information systems and others, directly to the organization. It enables cybersecurity teams, ethical hackers, and others to report vulnerabilities before they are generally known and in a way that is legal, transparent and in a positive interest of everyone. Security-conscious organizations publish Vulnerability Disclosure Policies containing clear steps and measures for reporting cybersecurity vulnerabilities, including the information that must be included, the contact parties, and so on.

A vulnerability disclosure process begins with an investigation of vulnerabilities by cybersecurity teams, researchers or ethical hackers, followed by establishing communication with the organization. Once the organization mitigates the vulnerability, a public disclosure is made. In this sense, the steps in the vulnerability disclosure process include [28]:

- **Discovery of Flaws:** Cybersecurity teams, researchers or ethical hackers discover vulnerabilities throughout manual testing or automated scanning tools, and try to understand the exploitability and impact of the vulnerability by verifying it in a controlled environment. Based on their investigation, they prepare a report and communicate the vulnerability to the organization.
- **Communication with the Organization:** To disclose a vulnerability, cybersecurity teams, researchers or ethical hackers try to identify if there is a Vulnerability

Disclosure Program (VDP) or a bug bounty program in place, to avoid any legal repercussions. A bug bounty program is a program advertised by an organization that places rewards on the discovery of vulnerabilities in, for example, software applications or web services. In other cases, communication can be established through intermediaries.
- **Internal Investigation and Mitigation**: The organization conducts an internal investigation to confirm the vulnerability and to understand its business impact. A patch or fix is prioritized and developed to mitigate the vulnerability and the cybersecurity team, researcher or ethical hacker is contacted to negotiate the time for cyber risk mitigation.
- **Public Disclosure:** Once the patch or fix is deployed, the organization may release the details to the public, in coordination with the cybersecurity team, researcher or ethical hacker. The discoverer may also be publicly recognized and compensated based on the agreed-upon terms.

Thus, a good vulnerability report should [29]:
- Describe the vulnerability precisely where it was discovered, and its real impact.
- Offer a detailed description of the steps needed to reproduce the vulnerability.
- Include one vulnerability per report (unless in an attack chain).
- Don't report automated scanner results without proof of exploitability.

Bug bounty program and penetration testing are effective ways to ensure organization's cybersecurity. Bug bounty is a crowdsourcing initiative that rewards individuals like ethical hackers and cybersecurity researchers for discovering and reporting vulnerabilities and bugs in software. Both programs have certain differences to be considered. The following characteristics distinguish the two methods, as illustrated in Table 2.11 [29]:

From Table 2.11 it can be seen that bug bounty programs and penetration tests are not in competition but complement each other.

Given this fact, a cybersecurity framework is required that guide cyber risk reduction. There are several frameworks available, ranging from NIST Cybersecurity Framework to ISO/IEC 27001-compliant ISMS. Furthermore, the development of the ISMS must also meet the cybersecurity requirements of NIS2. To deal with this, ISO/IEC 27000 [30] enables to describe the overview and the respective vocabulary of ISMS, referencing to the ISMS family of standards (including ISO/IEC 27003, ISO/IEC 26004, and ISO/IEC 27005), with the related terms and definitions:

- ISO/IEC 27003: Information technology, security techniques, Information Security Management Systems (ISMS), Guidance [31],
- ISO/IEC 27004: Information technology, security techniques, information security management, monitoring, measurement, analysis and evaluation [32],
- ISO/IEC 27005: Information security, cybersecurity and privacy protection, Guidance on managing information security risks [33].

These standards specify the requirements for establishing, developing, implementing, operating, monitoring, reviewing, maintaining and improving an ISMS within

Table 2.11 Characteristics of Bug-Bounty Program and Penetration Test

Bug-Bounty Program	Penetration Test
Continuously helps identify vulnerabilities, but testing activities are not guaranteed	Takes a snapshot of the cybersecurity situation
Report vulnerabilities, which must be checked and verified by the company for their correctness and novelty. The quality of the reported findings can vary greatly	Result is a comprehensive final report in which the vulnerabilities, including risk assessment and tailor-made proposals for action, are listed in a manner appropriate to the target group
Not clearly definable from a financial perspective. Organization must weigh attractiveness against cost. High rewards for vulnerabilities have a direct impact on tester interest. The expenses for the company are variable depending on the number of vulnerabilities found	Have a clearly defined price according to the infrastructure and individual specifications
Limited test duration not apply to a bug bounty program. Organization can take advantage of this feature strategically. If organization sets the scope on a small, particularly critical part of an application with a high reward for a vulnerability found, this leads to numerous test activities whose test time is not limited. This means that critical part can be examined in particular depth	Has a limited duration, during which the subject of the test is examined by experts

the needs of the organization's overall business cyber risk posture. In this context, organization shall do the following [30]: Define the scope and boundaries of ISMS in terms of the characteristics of the organizations business, its assets, technology, location, including details of and justification for any exclusions from the scope that

- includes a framework for setting objectives and establishes an overall sense of direction and principles for action to network and information security,
- considers organizations business and legal or regulatory requirements such as NIS2 [34], NIST CSF [35], MITRE ATTACK (MITRE ATT&CK) [16] and others, as well as contractual cybersecurity obligations,
- aligns with the organization's strategic cybersecurity risk-management within the establishment and maintenance of the ISMS will take place,
- establishes criteria against which cyber threat risk incidents be evaluated,
- has been approved by management.

Techniques for the management of network and information security require to identify the potential cyber risks.

Annex 14 of ISO/IEC 27001 refers to network and information system acquisition, development and maintenance based on several sub-and sub-sub-sections. Some of which, relevant to understand the whole scope, described in [36].

Annex 14.1 is about cybersecurity requirements of information systems, to ensure that information cybersecurity is an integral part of information systems across the entire lifecycle. Given that fact, in any new system development or change to existing systems, it is important to know what are the business requirements for cybersecurity controls, by undergoing a risk assessment. This should be done prior to the selection of or commencement of developing a solution.

Measures to Article 21.2.f: Policies and Procedures to Access the Effectiveness of Cybersecurity Risk-Management Measures

Cybersecurity risk-management measures are important because they enable a business to assess its current cybersecurity risk profile. Cybersecurity risk-management is the process identifying organization's digital assets, reviewing existing cybersecurity measures, and implementing solutions to either continue what works successful or to mitigate cyber risks that may pose cyber threat risk incidents to organizations business. Cybersecurity risk-management measures inform about decisions the cybersecurity organization team must enable moving forward in order to reduce the level of cyber risks and address vulnerabilities.

This subparagraph of Article 21 outlines the processes of identifying, prioritizing, managing and monitoring cyber risks to network and information system technology and information assets that focus on the evaluation and the enhancement of the effectiveness of established process(es). Consequently, organizations must carry out periodic reviews of their cybersecurity policies and procedures, whether they are continuously serving as vital components within their cybersecurity framework. Because of that, these assessments aim at identifying potential vulnerabilities, gaps, or outdated practices, using a cybersecurity risk assessment process that define cyber risks, identifying vulnerabilities and determining cyber risks likelihood and cyber risks impact to set up cybersecurity rules and controls for monitoring and review cybersecurity effectiveness to protect organizations data, networks, information systems and others [37]. For that reason, cybersecurity risk-management measures must proportionate to risk, size, cost, impact and severity of cyber risks and consider the state-of-the-art, as well as international standards, like ISO/IEC 27001:2022, Clause 9.1, to ensure the effective operation of risk-management and to implement appropriate analysis and evaluation of measures. Consequently, the status of cyber assets changes due to the dynamic cybersecurity environment of asset vulnerabilities, cyber risks, and recovery, the risk-management measures in assurance policy and risk-management enable decision makers to cope with the real-time assessment of the assurance level. For that reason, the organization must determine [38]: (a) what needs to be monitored and measured, including cybersecurity processes and controls; (b) the methods for monitoring, measurement, analysis and evaluation, as applicable, to ensure valid results. The methods selected should generate comparable and reproducible results to be considered valid; (c) when the monitoring and measuring shall be performed; (d) who shall monitor and measure; (e) when the results from monitoring and measurement shall be analyzed and evaluated; (f) who shall analyze and evaluate these results. Documented information shall be available as evidence of the results and continually evolve to follow the continual improvement processes in cybersecurity. In this regard, Threat Intelligence (TI) lead to changes to what to monitor and how to respond. In this sense, the organization shall evaluate the cybersecurity performance and effectiveness of the cybersecurity risk-management system. Against this background, organizations should try to stay ahead of the swiftly-evolving cybersecurity risk landscape, by conducting regular evaluations. Accordingly, establishing metrics and key performance indicators (KPIs) for cybersecurity controls are essential for continuous improvement of organization's cybersecurity risk-management procedures. This requires a proactive approach that enable protective and resiliency mechanisms at an early development stage, providing increased assurance that the security features, practices, procedures, and architecture of network and information systems are strong enough to defend or mitigate all known operational cyber risks. A best practice example for effective risk-management in cybersecurity is given in [39], which illustrate the several steps required:

- **Risk Assessment and Identification**:
  - Regularly assess of organization's assets, including digital data, applications, hardware and software, as well as customers databases, server infrastructure and others,
  - Identify potential cyber risks and vulnerabilities specific to the industrial infrastructure and technology environment,
  - Prioritize criticality of assets in the sense of business operations.

- **Quantitative and Qualitative Risk Analysis**:
  - Conduct quantitative and qualitative cyber risk assessments,
  - Quantitative analysis involves assigning values to cyber risks like calculating financial impact of cyber risks, reputation, probability, and others,
  - Qualitative analysis involves evaluating cyber risks based on subjective factors like reputational damage.

- **Recognized Risk-Management Framework Implementation**:
  - Adopt a recognized risk-management framework like NIST CSF 2.0, ISO/IEC 27001, MITRE Enable, and others
  - These frameworks provide structured guidelines for identifying, assessing, and mitigating cyber risks.

- **Regular Cybersecurity Audits and Testing:**
  - Conduct regular cybersecurity audits and penetration testing to uncover vulnerabilities by contracting ethical hackers,
  - Use automated tools and engage ethical hackers to simulate real-world cyber threat risk incidents.

- **Effective Risk Mitigation**:
  - Develop a policy and procedure for a cyber risk mitigation strategy that combines technology-based cybersecurity controls like firewalls, intrusion detection systems, employee cybersecurity situational awareness training, and others.
  - Prioritize high-risk areas for mitigation efforts.

- **Employee Training and Cybersecurity Situational Awareness (CSA):**
  - Educate employees on cybersecurity best practices like recognizing phishing attempts and importance of strong password management as well as their role in acting in risk-management,

- Promote a cybersecurity-conscious organization-wide culture where employees understand the potential impact of their actions taken.
- **Incident Response Plan**:
  - Create a comprehensive incident response plan that designates specific roles and responsibilities for cybersecurity team members and outlines steps to be taken in a cyber threat risk incident,
  - Regularly test and update this plan to ensure its effectiveness.
- **Continuous Monitoring**:
  - Implement continuous monitoring to detect unusual (anomalous) activities or potential cyber risks in real-time,
  - Use Security Information and Event Management (SIEM) solutions for effective monitoring.
- **Data Encryption and Access Controls**:
  - Encrypt sensitive customer data at rest and in transit, stored on organizations servers, and limit access to authorized personnel only,
  - Enforce strict access controls to ensure that only authorized personnel can access critical systems and digital data.
- **Patch Management**:
  - Keep all software, operating systems, and applications up to date with security patches to address known vulnerabilities,
  - Vulnerabilities in outdated software are often exploited by cybercriminals.
- **Backup and Disaster Recovery:**
  - Regularly back up critical data, networks and information systems by determining Recovery Tim Objective (**RTO**)and Recovery Point Objective (**RPO**) to minimize potential downtimes,
  - Develop a disaster recovery plan determining **RTO** and **RPO** to ensure Business Continuity (BC) in case of a cyber threat risk incident.
- **Third-party Risk-Management:**
  - Assess the cyber security practices of third-party vendors and managed service providers like cloud service providers, to ensure they align with the organization's cybersecurity standards, and contractual obligations.
- **Regular Cybersecurity Situational Awareness Training**:
  - Continuously educate employees about emerging cyber risks and evolving best practices, for example, simulated phishing exercises to educate employees about potential cyber threat risk incidents.
- **Regulatory Compliance**:
  - Stay informed about industry-specific regulations and compliance requirements like Network and Information Security 2 Directive (NIS2), Digital Operational Resilience Act (DORA), General Regulation Data Protection

(GRDP) and others, by regularly reviewing and updating organizations data protection policies,
- Align cybersecurity practices with relevant compliance standards like NIST CSF 2.0, ISO/IEC 27001, MITRE ATTAand others.

- **Board and Executive Involvement**:
  - Ensure that cybersecurity is a board-level concern,
  - Senior leadership should be actively involved in risk-management decisions and strategies.
- **Incident Documentation and Post-Incident Analysis**:
  - Thoroughly document all cyber threat risk incidents, including their causes and responses and the actions taken, and conduct a post-incident analysis to identify areas for improvements.
- **Cybersecurity Insurance:**
  - Consider cybersecurity insurance coverage to support mitigate financial losses in the event of cyber threat risk incidents,
  - Ensure that the policy covers the organization's specific risks.
- **Regular Updates and Adaptation**:
  - Cyber threat risk incidents evolve rapidly, so regularly update the risk-management strategy to adapt to new challenges,
  - Stay informed about emerging cyber threat risk incidents and vulnerabilities within the cybersecurity landscape.

As a conclusion a cybersecurity risk-management plan should be in accordance with the ISO/IEC 27001:2022 or the NIST CSF 2.0 risk-management standards, and other national or international cybersecurity initiatives, to achieve a high cybersecurity maturity level. The cybersecurity maturity level refers to an organization's security position relative to its cyber risk environment and tolerances. The cyber risk scenarios may vary greatly according to the different organizational environments, as each organization has its own cybersecurity situational aware and cyber risk culture. Consequently, NIST CSF 2.0 maturity levels are vital for organizations to gauge and enhance cybersecurity posture and strengths. The distinct levels of maturity represent a degree of maturity and capability in managing cyber risks. For example, at maturity level 1, the organization needs a structured cybersecurity risk-management process, while at maturity level 2, the organization have adopted a cyber risk-informed cybersecurity approach. At maturity level 3, the organization has established a standardized approach to cybersecurity risk-management, and maturity level 4 is the highest, which indicates that the organization has a proactive cybersecurity posture (see also Chap. 3). Especially, when cyber threat risk incidents are becoming increasingly sophisticated in their execution, NIST CSF 2.0 with the highest maturity level in place enable organizations can assess cybersecurity capabilities and continue to make informed improvements to their network and information systems.

Beside NIST CSF 2.0 [9], the NIST Special Publication 800-53 [17] serve different purposes but both can work together, because some overlap between them exist. NIST CSF 2.0 offers a broader cybersecurity structure, while NIST SP 800-53 provides a more specific cybersecurity control guidance.

In ISO/IEC 27001Annex A contains a set of cybersecurity controls that cover various aspects of network and information security for ensuring that organizations implement and maintains an effective ISMS, aligned with ISO/IEC 27001. It enables the organization focus on relevant cybersecurity controls, facilitates decision-making regarding the implementation of cybersecurity measures, and provides a clear reference for internal and external parties assessing the organization's information cybersecurity posture.

Measures to Article 21.2.g: Basic Cyber Hygiene Practices and Cybersecurity Training

The idea behind cyber hygiene is to create a structured environment that reduces cyber risks of external and internal contamination without having to consistently spend lots of IT effort on these processes. Thus, cyber hygiene can be introduced as set of practices and procedures that organizations use to maintain the health and cybersecurity resilience of their network and information systems, devices, and digital data for cyber risk assessment and identification to cyber risk-management of the most common and pervasive cyber threat risk incidents, faced by organizations, with regard to specific controls [40]:

- Identify and prioritize whether organizational assets like technologies, products, services, hardware, software and others be effective cyber-secured,
- Identify and prioritize cyber risks to organization's operations, including applications, functions, mission, reputation and others, and respond to identified cyber threat risk incidents, and continuously monitor to minimize the probability of occurrence and impact,
- Establish cyber threat risk incident response procedures, and if required, external partner(s) for handling disruption,
- Conduct Cybersecurity Situational Awareness (CSA) activities, to ensure personnel and partners are provided ongoing to CSA education to be adequately trained to perform their network and information systems cybersecurity-related duties and responsibilities,
- Ensure network and information system cybersecurity by consistently and appropriately configured network and information systems and continuous monitoring,
- Access need to control through least privilege, determine the validity period maintaining user access accounts to ensure user access information is current and accurate, and implement multiple authentication factors,
- Manage technology changes consistently by using standardized and cybersecure configurations, regularly validated and refreshed to update them with regard to recent cyber threat risk incidents, vulnerabilities, and attack surface vectors,

- Implement controls to protect and recover digital data against internal and/or external attack surface vectors, considering **RTO** and **RPO** measures,
- Prevent and continuously monitor malware exposures, e.g., through analysis by event log servers,
- Manage cyber risks associated with suppliers and external dependencies by processes to manage cyber risks and vulnerabilities that may result from supplier and third-party dependencies throughout the lifecycle of those relationships,
- Perform cyber risk and vulnerability information based on their probability, exploitability, and potential impact. Mitigate highest cyber risks and vulnerabilities by using monitoring, expected controls and patching timelines based on the cyber risk rating level and remediation.

However, not only organizations are affected by cyber risks and vulnerabilities, also human users are facing cyber risk or be a big cyber risk gap in cybersecurity that often is not given enough attention. Cyber threat risk incidents based on human errors are manifold, e.g. uploading sensitive data from an unauthorized system, using unsecure tools after changing a system configuration, coercion, while pressure exerted by a third party on an employee that enable them doing things they don't do under normal circumstances, and others. Another human based cyber threat risk incident is compromising, when employees clicks on an E-Mail attachment that access a malicious website, or using a USB stick with advertising text that installs malware on the employee's computer, and others. Cybercriminals exploit these possibilities to gain access to internal network and information systems. But also, insider cyber risks are human risks, where a person with legitimate access and company knowledge attempts blackmail or revenge. Beside this, the combination of coercion and compromising is a human cyber risk factor. Therefore, employee's CSA is an essential issue as well as remedial measures, including cybersecurity training to raise CSA, and should become a legitimate standard procedure in cyber hygiene and cybersecurity training.

Beside this, organizations cybersecurity culture also should have and live an error culture, to avoid that cyber threat risk incidents are not communicated. The background is that employees may be afraid about potential personal criticism and sanctions as a result of an error in their daily activity. Consequently, organizations employee's CSA is essential and should be investigated by using test methods that refers to targeted and/or pre-coordinated cyber threat risk incidents, as well as potential vulnerabilities in network and information systems, which can be executed automated or manually. Doing so, consent and knowledge are required, as some of the cyber threat risk incident methods used by a tester constitute cybercriminal offenses under the law. Given that fact, the aim of testing is both, control employees actual CSA, as well as uncover organizations vulnerabilities to provide concrete recommendations for action taken to eliminate deficiencies, e.g., closing existing cybersecurity gaps to increase the cybersecurity maturity level.

In addition to well-known test methods such as white box, black box and grey box audits, social engineering audits, and insider test are common procedures. Therefore, a preliminary discussion between IT cybersecurity experts commissioned to carry out tests, and organizations cybersecurity managers is necessary, defining aim and scope of the testss. The objectives can address to a wide variety of directions, and therefore must be planned and described precisely. In addition to the scope, the limits of action must be defined too. The methods and tools the tester use to check are the same as those used by cybercriminals.

- **Social Engineering Audit/Test:** Carried out to test employee's behavior, because cyber attackers no longer just use malware to penetrate organizations networks. Social engineering is not only one of the most efficient, but also one of the most dangerous methods of cybercrime, since employees are a major cyber risk group in organizations. Above all, the lack of CSA among employees of cyber threat risk incident variants, carried out via phishing, telephone or personal contact, let the method be popular for cybercriminals. Against this background, social engineering audits serves two purposes: On the one hand, checking CSA of employees, and on the other hand, uncover vulnerabilities and use as part of an analysis to develop an appropriate CSA concept. Evaluating compliance with organizations internal guidelines or the effectiveness of cybersecurity measures, it is often useful to combine an insider attack simulation with social engineering elements in order to test compliance with internal guidelines and effectiveness of cybersecurity measures.
- **Insider Test:** Vulnerabilities in cybersecurity also be available by means of by insider cyber risks. However, insider perpetrators do not necessarily have to be own employees, because also an external service provider or hijacked internal PCs can lead to inside perpetrator attacks. Since Small and Medium-Sized (SME) organizations in particular often hold the opinion that cybersecurity dangers come from outside, their systems are mostly secured against the outside, and often internal network and information systems is poorly secured. This means that in internal networks, often poorly protection mechanisms, undocumented systems and authorization assignments can be found.
- **Inside Perpetrators:** Divided into two groups: Perpetrators who act in a targeted manner and deliberately want to harm organizations, and perpetrators that will not recognize themselves as perpetrators, but rather acts unconsciously and carelessly. Thus, perpetrators try to enable that employees open anonymous emails, use unauthorized software or download of unknown attachments and others, and the unsuspecting employee has opened the door to cyber-attacks. Because, if cyber attacker has access to organizations systems and networks, they can calmly analyze the structure, uncover weaknesses and find a way to cause maximum damage with little effort. In the role of an insider, cyber attacker can intrude malware and accesses sensitive digital data, confidential information or asset resources. Given that fact, an insider test (in-test) is an authorized simulated cyber threat risk incident, performed on a computer system and standard user access of the organization under test, to intrude areas of the organization under test, expanding authorizations as far as possible. To achieve a real-world test outcome, testers checks when the organizations IT department becomes aware of the activities of the in-test intruder, whether existing security processes are suf-

ficiently effective and whether all cybersecurity standards are applied by all employees. The in-test results afterwards can be used to derive strategic recommendations that reliably eliminate identified vulnerabilities in the future.
- **Penetration Testing (pen-test):** Authorized simulated cyber-attack performed on computer systems to evaluate its cybersecurity enabled a real-world test outcome. Penetration testers use the same organizations tools, techniques and processes as probably cyber-attackers, to find and demonstrate the cyber-attack impacts of vulnerabilities/weaknesses in organizations processes and systems, as indicated in Table 2.8. Hence, penetration tests usually simulate a variety of potential cyber-attacks that could threaten organizations trustability and business assets. Furthermore, penetration tests also examine whether organization's processes and systems are robust to withstand potentially cyber-attacks from authenticated and unauthenticated positions, as well as the wide range of systems and roles in organizations. With the right scope, the penetration tester can dive into any critical and crucial area of systems [41]. However, it should be noted that a penetration test only shows uncertainties and is not a proof of comprehensive cybersecurity of organizations. Some pros and cons of penetration tests are summarized in Table 2.12.

The phases of successful planning and execution of penetration tests are [41, 42]:

- **Information Gathering:** Successful planning and execution of penetration tests requires outlining test logistic, test expectations and legal implications, as well as the objectives and goals a pen test customer would like to achieve. Information gathering requires rally penetration test strategy, which fit best for the organization under test, be identified to perform the pen test successful,
- **Reconnaissance:** After agreeing the type of penetration test with the pen test customer, the pen-tester have varying degrees of information about the pen test customers organization to identify critical information to uncover

Table 2.12 Pros and cons of penetration testing

Pros of penetration testing	Cons of penetration testing
Find vulnerabilities/weaknesses in upstream security assurance, list discovered vulnerabilities/weaknesses in order of criticality & priority of organization under test such as: • Password vulnerabilities/weaknesses • OT & endpoint application vulnerabilities/ weaknesses • Vulnerabilities/weaknesses misconfiguration • Injection to vulnerabilities/weaknesses • Cross-site Scripting (XXS) attacks	Labor-intensive, cumbersome and costly
Locate known & unknown software flaws & security vulnerabilities/weaknesses, including small ones that themselves won't raise much concern but could cause harm as part of complex attack pattern	Does not comprehensively prevent bugs & flaws from making their way into production
Can attach any system, mimicking how most malicious hackers would behave, simulating as close as possible a real-world adversary attack	

vulnerabilities/weaknesses and entry points in the customers digital environment. Thanks to this information the pen-tester uses an exhaustive checklist to find open entry points, vulnerabilities/weaknesses within the pen test customer organization. The Open-Source INTelligence (OSINT) framework provides a plethora of details for open information sources [43],

- **Discover Cybersecurity Threat Modeling and Vulnerability/Weakness Identification:** During this phase, the pen-tester targets, discovers, targets and maps the attack surface vectors. Any information gathered is used to decide about the cyber threat risk incident method in the penetration testing. The common areas the pen-tester map and discover include [41, 42],

  – Business Assets: Identify and categorize high value assets,

    Employee data,
    Customer data,
    Technical data,

  – Cyber Risks: Identify and categorize internal and external cyber risks,

    Internal cyber risks: Management, employees, vendors, and other,
    External cyber risks: Ports, Networks, Protocols, Web Applications, Network traffic, and other.

  – Pen-tester often use a vulnerability scanner to completely discover weaknesses. Thereafter, the pen-tester validate the exploitability of the vulnerability/weakness. The list of vulnerabilities/weaknesses is shared at the end of the penetration testing phase for the reporting phase [41, 42].

- **Exploitation**: With the map of probable vulnerabilities/weaknesses and entry points, the pen-tester begins with testing the exploits found so far in networks, applications, data, network and information systems and others of the pen test customer. The goal of the pen-tester is to really analyze how far he can intrude into the organizational environ-ment, to identify high-value targets, and avoid any detection, but only so far as agreed with the pen-test customer organization. Some standard exploit tactics include [41, 42],

  – Web Application Attacks: Attempt by cybercriminals to exploit vulnerabilities in web applications or mobile apps created during the software development process, with the goal disrupting business or gaining access to organizations IT ecosystems,
  – Network-based Attacks: Attempt by cybercriminals to breach or disrupt regular organizations network operations targeting data transmitted across networks, intended to steal information, wreak havoc, or disrupt services,
  – Memory-based Attacks: Attempt by cybercriminals targeting organizations computer's memory, including RAM, in order to exploit the system,
  – Wi-Fi-Attacks: Attempt by cybercriminals to exploit vulnerabilities of wireless communication systems, including Wi-Fi networks, mobile data networks, and Bluetooth connections.

- Zero-Day Angle Attacks: Attempt by cybercriminals taking advantage of cyber risks based on unpatched cybersecurity vulnerability, unknown to software, hardware or firmware developer.
- And others.

- **Pen-tester review and document how vulnerabilities/weaknesses are exploited and technics and tactics to obtain access to high-value targets. Finally, pen-tester explain the results from exploits on high-value targets such as:**

  - Post-Exploitation, Risk Analysis and Recommendation: Pen-tester document the methods used to gain access to pen-test organization's valuable assets. Pen-tester also be able to determine the value of compromised processes, systems and others and any value associated with sensitive data captured and provide recommendations on how to remediate the vulnerabilities/weaknesses within the pen test customer organization's environment for fixing cybersecurity holes and vulnerabilities/weaknesses.
  - Finally, one important activity is required once the penetration testing recommendations are completed, the pen-tester should clean up the pen test customers environment, reconfigure any access obtained to penetrate the pen test customers environment, and prevent future unauthorized access into the pen test customers organizational environment. Typical cleanup activities are [41, 42]:

    Removing any executables, scripts, and temporary files from compromised systems
    Reconfiguring settings back to the original parameters prior to the pen test
    Eliminating any rootkits installed in the customers environment
    Removing any user accounts created to connect to the compromised system
    And others.

  - Written Recommendations: From the pen-tester with regard to the findings and detailed explanations that offer insights and opportunities to significantly improve the cybersecurity posture of organizations processes, systems, applications and others. However, the report must explain how entry points were discovered from the OSINT and Discovering phase, as well as how to remediate the cybersecurity issues found in the exploitation phase. Finally, the report should also provide a roadmap of recommendations, mapping the major vulnerabilities /weaknesses to a defines timeline, as shown in Table 2.13, after [41, 44].

Pentesting ensures both cybersecurity coverage and compliance by uncovering critical vulnerabilities in organizations and educating IT- and engineering team to enhance cybersecurity best practices. However, Pentest-as-a-Service (PTaaS) redefines pentesting, offering flexibility, speed, and depth needed in modern, complex

Table 2.13 Mapping tasks to a defined timeline

Timeline 1–3 Months	
Tasks	To do's
Create Remediation Strategy	• Leverage results fund within pen test to create a full remediation strategy • Assessment report provide basis for this action; must be formalized & approved by CLIENT Security Team
Create Information Security Task Force	• Gain better tradition in remediation & security onboarding process. CLIENT create a specific ISEC council to aid in remediation and adequately involve each individual team • Council should consist of management of each individual business unit • …..
Begin Security Project Planning	• Assign executive owners of security for CLIENT • …..
Prioritize Remediation Events	• Leverage results found within pen test to gain understanding tasks needed be performed in order to resolve the risk(s) discovered • Assign priority listing to remediation tasks that will provide highest level of impact & largest reduction of discovered risk(s) • Start process with server patching to gain quick increases in environment security
Patch Services	• Specific things to be fixed/how …. • ……
Harden Servers	• ….. • …..

digital transformed ecosystems. Unlike traditional pentesting, PTaaS aligns with modern cybersecurity demands through a SaaS-based model, faster start times, and the power of a diverse, vetted cybersecurity community, as introduced by hackerone [45], to achieve cyber resilience, because it requires continuous adaptation to new cyber threat risk incidents. Regulations like the EU's Digital Operational Resilience Act (DORA) highlight this shift from compliance to constant readiness. PTaaS addresses this by aligning with continuous Cyber Threat risk incident Exposure Management (CTEM) and matching testing to re-al-world cyber attacker tactics to adapt defenses swiftly [45].

The foregoing best practice example for effective cyber hygiene in organizations cybersecurity is an essential requirement, but individuals also a minor role in maintaining organizations cybersecurity that requires specific settings to employees, partners, customers and others [46],

- **Use Strong Password**:
  – Creating a password means avoid short words and very simple numbers, such as first and last name, license plate, or birthday, as these are easy combinations to be hacked. Use uppercase and lowercase letters, as well as symbols with min of 8 characters, to create a strong, more elaborate password with elements that make it difficult to disclose.

- **Change Passwords Frequently**:
  - Personal data may be at high cyber risk if the same password is used across several websites or services. If a keyword is hacked, criminals will look for other accounts of that same user. So, change credentials regularly, which is what password management can support to do.
- **Don't Open Suspicious E-Mails**:
  - Receiving a strange email e.g. from the bank! Don't open the message, as it may be the target of a scam. Cybercriminals can gain access to critical information and even use the organizations computer to break into the organizations internal network.
- **Avoid Downloads from Unknown E-Mails**:
  - Software is often the target of viruses and scams by cybercriminals. Avoid downloading software, as they can steal data and passwords from organizations computer(s). This also applies to browser extensions (applications that add extra functions to programs). Hence, check user ratings and comments about the app before downloading.

In contrast, cyber hygiene best practices for Small and Medium Enterprises (SME) in general need some special considerations like [46].

- **Back-up Organizations Data:** Practice of copying data from a primary to a secondary location, to protect it in case of a disaster, or malicious action to avoid disrupt business operations.
- **Grant Limited Access Rights**: Enables employees to access a specific content item of the organization when they do not have permissions to open any other items in the organizations data files.
- **User Multi-Factor Authentication (MFA):** Multi-step account login process that requires users to provide two or more verification factors to gain access to an organizations resource. MFA is a core component of a strong Identity and Access Management (IAM) policy.
- **Know Organizations Legal Obligations:** Denotes any requirement generated through the common law that have to be known to perform and accomplish it.
- **Provide Cybersecurity Training to Employees:** *Employees* are the first line of defense against cyber threat risk incidents. Thus, employees need to understand the importance of cybersecurity and need training how to identify potential cyber threat risk incidents and respond appropriately. Cybersecurity awareness training also provides employees with the knowledge and skills needed to recognize, report, and prevent security incidents.
- **Update and Patch Software and Systems:** Patches are software and Operating System (OS) updates that address cybersecurity vulnerabilities within a program or product. Software vendors may choose to release updates to fix performance bugs, as well as to provide enhanced cybersecurity features. Manual updates require to visit the vendor's website to download and install software files, automatic updates require consent when installing or configuring the software.

- **Implement Email Cybersecurity:** Organizations must enhance their email cybersecurity posture by establishing policies and using tools to protect against malicious cyber threat risk incidents such as malware, spam, and phishing attacks.
- **Establish Policies:** Process by which organizations authorities create rules, regulations, and laws that guide the actions and decisions of organizations in cybersecurity.
- **Use Next-Generation Endpoint Cybersecurity Software:** Applies Artificial Intelligence (AI), Machine Learning (ML), and a tighter integration of network and device cybersecurity to provide more comprehensive and adaptive protection than traditional endpoint security solutions.
- **Buy Cyber Insurance:** Covers financial losses that organizations have as a result of ransomware attacks, data breaches and other cyber threat risk incidents.
- And others.

Measures to Article 21.2.h: Policies and Procedures Regarding the Use of Cryptography and, Where Appropriate, Encryption

This policy defines controls and related procedures where encryption and other cryptographic techniques are deployed to achieve information cybersecurity objectives, enabled by transfer of sensitive digital data throughout a cybersecure channel, whereby a cybersecure channel is an encrypted network or information system connection. Encryption is the process of protecting data by means of mathematical models to scramble in such a way that only the ones who have the key to unscramble it, can access it, which means that data encryption prevents unauthorized users from accessing organization sensitive digital data. Cryptography is the process of hiding or coding information so that only the person a message was intended for can read it. The difference between encryption and cryptography is that cryptography is the method of concealing messages with a secret code, while encryption is the way to encrypt and decrypt data. Various methods of encryption are available and generally built-into the application. With it, organizations should be aware of the data connection being used to transmit sensitive data and if encryption is enabled for that connection. From a general perspective encryption is required for:

- Transport of sensitive digital data, and files that need SSL (Secure Socket Layer) or SCP (Secure, Contain, Protect), to encrypt sensitive digital data for network file access of unencrypted files. SSL cybersecure internet connection by encrypting digital data sent between a website and a browser, or between two servers, preventing cybercriminals from seeing or stealing any digital data and information transferred,
- Access to sensitive digital data via a web site, web application or mobile app requires encryption for accessing sensitive digital data that need to use HTTPS (Hypertext Protocol Secure), the cybersecure web protocol, enabled by SSL

(Secure Socket Layer) and TLS (Transport Layer Security). TLS is an updated more secure version of SSL,
- Network traffic for remote access to a virtual desktop environment need encryption,
- Transport of sensitive digital data that is part of a database query SQL (Standard Query Language) or web service call need encryption,
- Privileged access to network or server equipment for system management purposes; requires SSH (Secure Shell or Secure Socket Shell) for encryption. SSH is a protocol for securely sending commands to a computer over an unsecured network. SSH uses cryptography to authenticate and encrypt connections between devices,
- And others.

Owing encryption is the process of converting data from plain text to a form that is not readable for those unauthorized, known as cipher-text. In encryption a key is used that controls the process of encryption and decryption. The policy lays out general requirements relating to cryptographic controls such as encryption and digital signatures, where these are needed to minimize unacceptable information, files and data risks that cannot be adequately mitigated through other means or for compliance reasons. Given that fact, cryptographic controls can be used to achieve different network and information cybersecurity objectives such as [47]:

- **Confidentiality:** Using encryption of information to protect sensitive or critical information, files and digital data, either stored or transmitted,
- **Integrity/Authenticity:** Using digital signature certificates or message authentication codes to verify authenticity or integrity of stored or transmitted sensitive or critical information, files and digital data,
- **Non-repudiation:** Using cryptographic techniques to provide evidence of the occurrence of a cyber risks,
- **Authentication:** Using cryptographic techniques to authenticate users and network and information system entities, requesting access or transacting with system users, entities and resources,
- **Encryption of sensitive or critical information, files and data:** Adds an extra layer of protection against probable cyber risks, rendering information, files and digital data unusable even if a potential cybercriminal manages to exfiltrate it,
- **End-to-End Encryption:** Improves information, files and digital data cybersecurity in remote access scenarios, where it shields the information, files and digital data from eavesdropping or Man-in-the-Middle (MITM) attacks.

Policy requirements for protecting encryption are under the control of end users. These requirements prevent unauthorized disclosure and subsequent fraudulent use. The protection methods outlined include operational and technical controls. If not done properly, encryption can lead to compromise and disclosure of private encryption to cybersecure sensitive digital data and compromise of the digital data. Although organizations must understand it's important to encryption certain documents and electronic communications, they may not be familiar with minimum mechanisms for encryption. Encryption mechanisms are in the focus of international standards that specifies a number of asymmetric ciphers, including identity-based ciphers, block ciphers, stream ciphers, and homomorphic encryption. ISO/IEC 27001 Annex A Controls and especially Annex 8.24 Use of Cryptography requires to define and manage rules associated with cryptography that in laymen's terms is encryption [48]. The purpose ISO/IEC 27001 Annex A 8.24 is a preventive control to ensure effective use of cryptography to protect confidentiality, authenticity or integrity of information according to business and information cybersecurity requirements, and taking into consideration legal, statutory, regulatory and contractual requirements related to cryptography.

Measures to Article 21.2.i: Human Resources Security, Access Control Policies and Asset Management

Human resources security, access control policies and management, and asset management refers to indispensable requirements in cybersecurity control. In this context, human resources cybersecurity is built of a set of processes designed to ensure that all employees, suppliers, and contractors of an organization are qualified for and understand their engagement/job tasks and responsibilities within the organization, as described in the ISO/IEC 27001 standard, and that access is revoked after the engagement is finished. Therefore, regulating which users (employees, suppliers, contractors) can access an organizations critical and crucial assets, digital data, and information resources or perform specific actions in an organization's digital environment, to ensure that only authorized users are granted access while preventing unauthorized users from gaining entry. This usually con-sists on the one hand an organizations clear onboarding and offboarding policy to prevent departed users can use their account furthermore after the engagement is finished, and on the other hand of authentication and authorization.

Authentication is basically a cybersecurity check to verify whether users are who they claim to be, which can be done with the combination of usernames (to identify the users) and passwords (to prove it's really them), before the user was granted access to organizations critical and crucial systems. Thus, authentication include different types of methods like [49]:

- Knowledge-based Authentication:
  Static Type: Users only required to select a security question and set the answer to the question.
  Dynamic Type: Authentic user is likely to know the answer to the random question as it is usually based on their records and history of activities or transactions.
- Possession-based Authentication: Examples include:
  Hardware Token: A physical device the user possesses and used to generate tokens (code) used to authenticate themselves.
  One-time Password (OTP): Set of keys generated for one-time use when accessing a system. Be sent to the user via text message (SMS) or email.
  Time-based One-Time Password (TOTP): Uses a special algorithm to generate a unique one-time use password. Algorithm uses the current time and a secret key that is shared between a user's device and the requested ones.

- Biometric Authentication: Biometric attributes such as fingerprint and facial recognition characteristics are more difficult to steal compared to text-based passwords.

Authorization determines what critical assets, digital data and information resources a user has access to and in what circumstances, such as the device he is on, the location, the role, and others. These facts are to determine whether or not a user have permission to access critical and crucial assets, data and information resources he is trying to access.

In addition, least-privilege access control policies like Zero-Trust Network Access (ZTNA), also known as the Software-Defined Perimeter (SDP), is a set of technologies and functionalities that enable cybersecure access to internal applications for remote users, and asset management, significantly reduce the cyber risk of damage caused by inadvertent error and dampen the impact of credential compromise by restricting the pool of resources accessible by a hijacked account.

In particular, network-based ZTNA is a straightforward way of cybersecuring older systems that do not support application-level access controls and Multi-Factor Authentication (MFA), an authentication method that requires the user to provide two or more verification factors to gain access to an organizations digital resource, and their configuration for this purpose would otherwise be difficult [50].

Against this background, human resources cybersecurity, access control policies and management that defines rules for keeping under control users access to organizations applications, data and information, and asset management are vital of any cybersecurity control in organizations. It ensures that all potential users have the right level of access to organizations essential resources and keeps everyone else out. In case of a cyber threat risk incident, this limits the attack vectors that cyber attackers can exploit. It keeps critical and confidential assets, information and digital data resources safe from being stolen, manipulated by malicious cybercriminals or unauthorized users and puts a lid on web-based cyber threat risk incidents. The NIS 2 directive mention access control in the same bucket as human resources cybersecurity and asset management, which makes sense as asset management is a prerequisite for access control, and human resources is an important player in enforcing access controls [50]. In this regard, ISO/IEC 27001 covers in Annex 7 the human resource security issue and in Annex 8 asset management.

The objective in ISO/IEC 27001 Annex 7 is to ensure that employees and contractors meet their responsibilities and are suitable for the roles for which they are considered. It also covers what happens when those people leave or change roles, and is an important part of the ISMS, especially to achieve ISO 27001 certification [51]. A serious control requires that all employees, suppliers, and contractors that use organizations network and information resources, act in accordance with the organizations policies and procedures. Thus, Annex 7.1 states:

- Ensure that organization's managers responsibility should also include requirements to ensure that they are responsible to understand information cybersecurity threat risk incidents, vulnerabilities and controls relevant to their job roles and receive regular training (as per ISO/IEC 27001 Annex 7.2.2),

- Ensure buy-in to proactive and adequate support for essential network and information cybersecurity policies and controls, and reinforce the requirements of terms and conditions of employment.

Thus, managers comply with a critical role in ensuring cybersecurity consciousness and conscientiousness throughout the organization and in developing an appropriate cybersecurity culture

In addition, ISO/IEC 27001 covers in Annex 7.2.2 information cybersecurity awareness, education and training. All employees, suppliers, and contractors must receive appropriate cybersecurity awareness education and training to satisfy their job well and cybersecure. Furthermore, they must obtain regular updates in organizations policies and procedures when they are changed too, along with a serious understanding of the applicable legislation that affects them in their role.

To establish this activity the cybersecurity team can partner with HR (Human Resources) or the training and development team to carry out skills, knowledge, competence and awareness assessments, to plan and implement a program of awareness, education and training throughout the employment lifecycle. Beside this, it is required to prove that training and compliance to auditors. Moreover, regular cybersecurity awareness trainings delivered to employees, suppliers and contractors is the key possibility to keep them updated in their daily work environment. ISO/IEC 27001 Annex A.7.3 is about termination and change of employment. The objective in this Annex is to protect the organization's interests as part of the process of changing and terminating employment.

Measures to Article 21.2.j: Use of Multi-Factor Authentication or Continuous Authentication Solutions, Secured Voice, Video and Text Communications and Secures Emergency Communication Systems Within the Entity, Where Appropriate

The issues listed in Article 21.2.j refers to indispensable requirements in cybersecurity control through authentication and cybersecure communication and emergency communication.

The primary focus of Multi-Factor Authentication (MFA) is providing higher cybersecurity through multiple factor authentication methods (see Article 21.2.h). However, this additional cybersecurity benefit comes at the cost of some user convenience, as users need to go through multiple steps to access their accounts. Cybersecurity benefits by MFA for organizations include:

- Increased protection from cyber risks. Thus, MFA enables to protect valuable assets, digital data and information resources,
- Enable safe remote work environments to keep the remote work environment flexible and agile,
- Defense in depth through multiple cybersecurity layers so that if one layer of defense is intentionally or accidentally compromised, secondary and tertiary layers (and so on) provide a backup, making sure that an organization is protected to the best possible level.

MFA be implemented on the application layer, for example, as part of SSO (Single Sign On), an identification method that enables users to log in to multiple applications and websites with one set of credentials, streamlining the authentication process for them, or on the network layer, during authentication on the SDP (Software-Defined Perimeter), a way to hide Internet-connected infrastructure (servers, routers, etc.) so that external parties and cybercriminals cannot see it, whether it is hosted on-premise or in the cloud. More in general, organizations that uses SDP are essentially draping a cloak of invisibility over their servers and other infrastructure resources, so that no one can see it from the outside; but authorized users can still access them.

In contrast, SSO prioritizes convenience by allowing users to access multiple accounts with a single login. While this can improve workflow and reduce time lost to password resets, it also means that if the SSO is compromised, all connected accounts are at cyber risk.

According to NIS2 Directive [52], Article 21.2.j, as a rule, medium-sized or larger social networking platforms have to take appropriate and proportionate cybersecurity risk-management measures, including the use of MFA or continuous authentication solutions, secured voice, video and text communications and secured emergency communication within the organization. An emergency communication system enables organizations to exchange relevant information in case of cybersecurity danger. Thus, a suitable emergency communications plan is required that is pivotal addressing the who, when, and what to communicate during a cybersecurity breach. This is required for stakeholders, media, governance, the general public, and others. In an age of sophisticated cybersecurity breaches, a clear communication pathway support to mitigate damage and preserve organizational reputation. Thus, a suitable emergency communication plan can prevent panic and halt the dissemination of false information. Therefore, most of the following proactive measures for practical reasons are required:

- List of internal contacts, e.g., organization's C-suite, managers, and employees,
- List of external contacts, e.g., customers, stakeholders, business partners, suppliers, investors, government agencies, and the media,
- Specific templates, e.g. call logs to capture media and individual inquiries, emergency contact directory, incident description report,
- Pre-written documents, e.g. press releases, initial announcements and follow-up statements,
- Setting guidelines for dealing with social media in an emergency, e.g., who can issue messages, which media, social messaging platforms can be used,
- Trained emergency communications team with knowledge of messaging systems/social media,
- Trained spokesperson and at least one deputy of the organization,
- Technology to quickly disseminate emergency information to employees, stakeholders, suppliers, customers, government agencies and other external entities,
- Binding policies on all aspects of emergency communications of the organization.

2.5 Chapter IV Cybersecurity Risk-Management Measures and Reporting Obligations

As a recently add on, the Commission had until 17 October 2024 to adopt implementing acts about the technical and the methodological requirements of the cybersecurity risk-management measures also with respect to social networking platforms. The work for these acts done in the framework of the NIS Cooperation Group [53].

In addition, the Commission introduced the Cyber Resilience Act (CRA) [54], which applies to products with digital elements, including standalone software and applications placed on the EU market and throughout their product life cycle. Manufacturers need to consider essential cybersecurity requirements, including that products ensure protection from unauthorized access by appropriate control mechanisms, including but not limited to authentication, identity or access management systems.

The CRA is a legislation that explicitly addresses OT components and critical infrastructures. CRA Annex 1 defines essential requirements that are subdivided in needs for cybersecurity product properties and for vulnerability handling. In this context, product properties refer to product development properties, because digital products must undergo a cyber threat risk incident analysis during development, they must not contain any exploitable vulnerabilities, which results in specific requirements, for example, cybersecurity by default configurations, access protection, CIA Triad, data economy, hardening, logging and monitoring, and the requirement that vulnerabilities must be remediable through security updates.

In important applications, vulnerability handling requires a SBOM (Software Bill of Materials) in a machine-readable format as mandatory need, because vulnerabilities must be fixed immediately and reported to ENISA that runs the EU Interinstitutional Services, the Cybersecurity Service for the Union Institutions, Bodies, Offices and Agencies (CERT-EU). SBOM is a key building block in software cybersecurity and software supply chain risk management, described as an inventory of all the components, including third-party libraries, artifacts, licenses, scripts, and package versions, as well as dependencies that make up a software application. As an inventory list, the SBOM provides transparency into all constituent parts of the software. Using an SBOM, organizations can understand what their applications rely on and identify vulnerabilities or license risks that may impact them. Thus, SBOM enables identify and avoid known vulnerabilities, quantify and manage licenses, identify cybersecurity and license compliance, and manage mitigation of vulnerabilities.

Article 21.2 Paragraphs 3–5: Cybersecurity Risk-Management Measures

3. Member States shall ensure that, when considering which measures referred to in paragraph 2, point (d), of this Article are appropriate, entities consider the vulnerabilities specific to each direct supplier and service provider and the overall quality of products and cybersecurity practices of their suppliers and service providers, including their secure development procedures. EU Member States shall also ensure that, when considering which measures referred to in that point

are appropriate, entities are required to consider the results of the coordinated security risk assessments of critical supply chains carried out in accordance with Article 22(1).
4. Member States shall ensure that an entity that finds that it does not comply with the measures provided for in paragraph 2 takes, without undue delay, all necessary, appropriate and proportionate corrective measures.
5. By 17 October 2024, the Commission shall adopt implementing acts laying down the technical and the methodological requirements of the measures referred to in paragraph 2 with regard to DNS service providers, TLD name registries, cloud computing service providers, data center service providers, content delivery network providers, managed service providers, managed security service providers, providers of online market places, of online search engines and of social networking services platforms, and trust service providers.

The Commission may adopt implementing acts laying down the technical and the methodological requirements, as well as sectoral requirements, as necessary, of the measures referred to in paragraph 2 with regard to essential and important entities other than those referred to in the first subparagraph of this paragraph.

When preparing the implementing acts referred to in the first and second subparagraphs of this paragraph, the Commission shall, to the extent possible, follow European and international standards, as well as relevant technical specifications. The Commission shall exchange advice and cooperate with the Cooperation Group and ENISA on the draft implementing acts in accordance with Article 14(4), point (e).

Those implementing acts shall be adopted in accordance with the examination procedure referred to in Article 39(2).

In this context, Article 21.2 Paragraphs 3–5 illustrate that the NIS2 Directive describes what needs to be achieved, although it doesn't prescribe how one must achieve it targets. For the subjects 21.2 a-j, international standards enable asset owners to implement the right set of control to cybersecure their operations as illustrated in Table 2.14.

A correspondence between NIS2 Article 21.2 and ISO/IEC 27001 Annexes is shown in Table 2.15 for completeness.

In [55] a guideline for time critical systems is introduced to map the NIS2 requirements against IEC 62443 cybersecurity requirements. The IEC 62443 series of standards define requirements and processes for implementing and maintaining electronically cybersecure Industrial Automation and Control Systems (IACS). IEC 62443 introduce a set best practices for cybersecurity and provide a way to assess the level of cybersecurity performance, applicable to asset owners and operators to safeguard IACS. The approach of IEC 62443 is holistic bridging the gap between OT (Operations Technology) and IT (Information Technology) as well as between process safety and cybersecurity, and offer a robust framework, which covers the topics of risk assessment, security policies, network architecture, access control, incident response, and security testing. The most relevant IEC 62443 standard for the EU NIS Directive is IEC 62443-2-1 security program requirements for IACS assets in organizations. Table 2.16 illustrate how IEC 62443 can guide implementing cybersecurity risk-management measures that NIS2 requires.

2.5 Chapter IV Cybersecurity Risk-Management Measures and Reporting Obligations

Table 2.14 Comparison NIS2 Article 21.2 with international Standards

Article 21.2	Measures	International Standards
a	Risk analysis and information security guidelines	NIST CSF, NIST 800-53, ISO/IEC 2700, ENISA Risk-Management/Risk-Analysis
b	Handling incidents	ISO /IEC 27036-2:2023, NIST SP 800-61
c	BCM: Backup Management BCM: Disaster Recovery BCM: Crisis Management	NIST SP 800-209, ISO/IEC 27040:2015 NIST SP 800-184, ISO/IEC 27031:2011, NIST SP 800-34, ISO/IEC 22301:2019
d	Supply chain security	NIST SP 800-16, ENISA Interoperable EU Risk-Management System
e	Security in NIS: Acquisition, development, maintenance, handling and disclosure of vulnerabilities	NIST SP 800-216, ISO/IEC 29147:2018, EU CRA NIST SP 800-100, -123, -137 NIST SP 800-216, ISO/IEC 29147:2018, ENISA Coordinated Vulnerability Disclosure B
f	Guidelines for assessing effectiveness of risk-management measures in cybersecurity	NIST SP 800-55, ENISA National Cyber-security Assessment Framework (NCAF) Tool, MITRE ATT&CK
g	Basic cyber hygiene, Cyber security training	NIST SP 800-53, NISTIR 7621, NIST 1800 NIST NICE Framework, ENISA ECSF
h	Guidelines for the use of cryptography, if necessary encryption	ISO/IEC 27001 family of standards NIST SP 800-175A and B
i	Security Human Resources Access control policies Asset management	NIST SP 800-50, ENISA's AR-in-a-Box, NIST SP 800-192, NISTIR 7316, NISTIR 7874, ISO/IEC 29146:2016 NIST SP-1800-6, ISO/IEC 19770
j	MFA or continuous authentication, secure voice, video and text communications Secure emergency communication systems	NIST SP 800-63B, ISO TR 29156:2015 NIST SP 800-45, NIST SP 800-177

Table 2.15 Mapping NIS2 Article 21 to ISO/IEC 27001 Annexes

NIS2 Article 21.2 Security Domains	ISO/IEC 27001 Annexes
Asset Management	A.8: Asset Management
Processing and Reporting of Incidents	A.16: Information Security Incident Management
Business Continuity (BC)	A.17: Information Security Aspects for BC
Cryptography and Encryption Techniques	A.10: Cryptography
Personnel Security	A.7: Personal Security
Access Control Policies	A.9: Access Control A.5: Information Security Policies
Security in the Supply Chain	A.15: Supplier Relationships
Security in System Acquisition, Development and Maintenance	A.14: System Acquisition, Development, and Maintenance
Zero-Trust Security (MFA)	A.9: Access Control

Table 2.16 Comparison between NIS2 Article 21.2 and IEC 62443-2-1

NIS2 Article 21.2	IEC 62443-2-1
a.	4.3.2.3 Organizing for security 4.3.2.6 Security policies and procedures 4.4.3 Review, improve and maintain the CSMS
b.	4.3.4.5 Incident planning and response
c.	4.3.2.5 Business continuity plan 4.3.4.5 Incident planning and response
d.	4.3.2.2 CSMS scope 4.3.2.3 Organizing for security 4.3.4.3 System development and maintenance 4.4.3 Review, improve and maintain the CSMS Also use IEC62443-2-4 Security program requirements for IACS service providers
e.	4.2.3 Risk identification, classification and assessment 4.3.3.4 Network segmentation 4.3.4.3 System development and maintenance
f.	4.3.2.6 Security policies and procedures 4.2.3 Risk identification, classification and assessment 4.4.2 Conformance 4.4.3 Review, improve, and maintain the CSMS
g.	4.3.2.4 Staff training and security awareness IEC62443-2-4 SP 01.01-03 Solution Staffing—Training
h.	4.3.4.3 System development and maintenance
i.	4.3.2.4 Staff training and security awareness 4.3.3.2 Personnel security 4.3.3.5 Access control—Account administration 4.3.3.6 Access control—Authentication 4.3.3.7 Access control—Authorization 4.3.4.4 Information and documentation management
j.	4.3.2.5 Business continuity plan 4.3.3.5 Access control—account administration 4.3.3.6 Access control—authentication

2.5.3 *Article 22 Union Level Coordinated Security Risk Assessments of Critical Supply Chains*

Article 22 Union Level Coordinated Security Risk Assessments of Critical Supply Chains Describes this topic in two paragraphs self-explanatory.

1. The Cooperation Group, in cooperation with the Commission and ENISA, may carry out coordinated security risk assessments of specific critical ICT services, ICT systems or ICT products supply chains, considering technical and, where relevant, non-technical risk factors.
2. The Commission, after consulting the Cooperation Group and ENISA, and, where necessary, relevant stakeholders, shall identify the specific critical ICT services, ICT systems or ICT products that may be subject to the coordinated security risk assessment referred to in paragraph 1.

The European Union Agency for Cybersecurity, ENISA, named in article 22, is the EUs agency dedicated to achieving a **high common level of cybersecurity across the EU.** Established in 2004 and strengthened by the EU Cybersecurity Act, ENISA contributes to EU cyber policy, enhances the trustworthiness of ICT products, services and processes with cybersecurity certification schemes, cooperates with EU Member States and EU bodies, and helps EU-Europe prepare for the cyber challenges of tomorrow. Through knowledge sharing, capacity building and awareness raising, the ENISA works together with its key stakeholders to strengthen trust in the connected economy, to boost resilience of the EU's infrastructure, and, ultimaly, to keep EU-Europe's society and citizens digitally secure [56].

2.5.4 Article 23 Reporting Obligations

Article 23 Reporting Obligations Describes this topic in 11 paragraphs that are partially self-explanatory in principle. However, a subsequently separate reflection to paragraph 4 considers in detail to essential and important sector organizations obligations.

1. Each Member State shall ensure that essential and important entities notify, without undue delay, its CSIRT or, where applicable, its competent authority in accordance with paragraph 4 of any incident that has a significant impact on the provision of their services as referred to in paragraph 3 (significant incident). Where appropriate, entities concerned shall notify, without undue delay, the recipients of their services of significant incidents that are likely to adversely affect the provision of those services. Each Member State shall ensure that those entities report, inter alia, any information enabling the CSIRT or, where applicable, the competent authority to determine any cross-border impact of the incident. The mere act of notification shall not subject the notifying entity to increased liability. Where the entities concerned notify the competent authority of a significant incident under the first subparagraph, the Member State shall ensure that that competent authority forwards the notification to the CSIRT upon receipt.

 In the case of a cross-border or cross-sectoral significant incident, Member States shall ensure that their single points of contact are provided in due time with relevant information notified in accordance with paragraph 4.

2. Where applicable, Member States shall ensure that essential and important entities communicate, without undue delay, to the recipients of their services that are potentially affected by a significant cyber threat any measures or remedies that those recipients are able to take in response to that threat. Where appropriate, the entities shall also inform those recipients of the significant cyber threat itself.

3. An incident shall be considered to be significant if:

   (a) it has caused or is capable of causing severe operational disruption of the services or financial loss for the entity concerned;
   (b) it has affected or is capable of affecting other natural or legal persons by causing considerable material or non-material damage.

4. Member States shall ensure that, for the purpose of notification under paragraph 1, the entities concerned submit to the CSIRT or, where applicable, the competent authority:

   (a) without undue delay and in any event within 24 h of becoming aware of the significant incident, an early warning, which, where applicable, shall indicate whether the significant incident is suspected of being caused by unlawful or malicious acts or could have a cross-border impact;
   (b) without undue delay and in any event within 72 h of becoming aware of the significant incident, an incident notification, which, where applicable, shall update the information referred to in point and indicate an initial assessment of the significant incident, including its severity and impact, as well as, where available, the indicators of compromise;
   (c) upon the request of a CSIRT or, where applicable, the competent authority, an intermediate report on relevant status updates;
   (d) a final report not later than 1 month after the submission of the incident notification under point (b), including the following:

      (i) a detailed description of the incident, including its severity and impact;
      (ii) the type of threat or root cause that is likely to have triggered the incident;
      (iii) applied and ongoing mitigation measures;
      (iv) where applicable, the cross-border impact of the incident;

   (e) in the event of an ongoing incident at the time of the submission of the final report referred to in point (d), Member States shall ensure that entities concerned provide a progress report at that time and a final report within 1 month of their handling of the incident.

   By way of derogation from the first subparagraph, point (b), a trust service provider shall, with regard to significant incidents that have an impact on the provision of its trust services, notify the CSIRT or, where applicable, the competent authority, without undue delay and in any event within 24 h of becoming aware of the significant incident.

5. The CSIRT or the competent authority shall provide, without undue delay and where possible within 24 h of receiving the early warning referred to in paragraph 4, point (a), a response to the notifying entity, including initial feedback on the significant incident and, upon request of the entity, guidance or operational advice on the implementation of possible mitigation measures. Where the CSIRT is not the initial recipient of the notification referred to in paragraph 1, the guidance shall be provided by the competent authority in cooperation with

the CSIRT. The CSIRT shall provide additional technical support if the entity concerned so requests. Where the significant incident is suspected to be of criminal nature, the CSIRT or the competent authority shall also provide guidance on reporting the significant incident to law enforcement authorities.

6. Where appropriate, and in particular where the significant incident concerns two or more Member States, the CSIRT, the competent authority or the single point of contact shall inform, without undue delay, the other affected Member States and ENISA of the significant incident. Such information shall include the type of information received in accordance with paragraph 4. In so doing, the CSIRT, the competent authority or the single point of contact shall, in accordance with Union or national law, preserve the entity's security and commercial interests as well as the confidentiality of the information provided.

7. Where public awareness is necessary to prevent a significant incident or to deal with an ongoing significant incident, or where disclosure of the significant incident is otherwise in the public interest, a Member State's CSIRT or, where applicable, its competent authority, and, where appropriate, the CSIRTs or the competent authorities of other Member States concerned, may, after consulting the entity concerned, inform the public about the significant incident or require the entity to do so.

8. At the request of the CSIRT or the competent authority, the single point of contact shall forward notifications received pursuant to paragraph 1 to the single points of contact of other affected Member States.

9. The single point of contact shall submit to ENISA every 3 months a summary report, including anonymized and aggregated data on significant incidents, incidents, cyber threats and near misses notified in accordance with paragraph 1 of this Article and with Article 30. In order to contribute to the provision of comparable information, ENISA may adopt technical guidance on the parameters of the information to be included in the summary report. ENISA shall inform the Cooperation Group and the CSIRTs network about its findings on notifications received every 6 months.

10. The CSIRTs or, where applicable, the competent authorities shall provide to the competent authorities under Directive (EU) 2022/2557 information about significant incidents, incidents, cyber threats and near misses notified in accordance with paragraph 1 of this Article and with Article 30 by entities identified as critical entities under Directive (EU) 2022/2557.

11. The Commission may adopt implementing acts further specifying the type of information, the format and the procedure of a notification submitted pursuant to paragraph 1 of this Article and to Article 30 and of a communication submitted pursuant to paragraph 2 of this Article.

By 17 October 2024, the Commission shall, with regard to DNS service providers, TLD name registries, cloud computing service providers, data center service providers, content delivery network providers, managed service providers, managed security service providers, as well as providers of online marketplaces, of online search engines and of social networking services platforms, adopt implementing acts further specifying the cases in which an incident shall be considered to be significant as referred to in paragraph 3. The Commission may adopt such implementing acts with regard to other essential and important entities.

The Commission shall exchange advice and cooperate with the Cooperation Group on the draft implementing acts referred to in the first and second subparagraphs of this paragraph in accordance with Article 14(4), point (e).

Those implementing acts shall be adopted in accordance with the examination procedure referred to in Article 39(2).

The reporting obligations in Article 23.4 focus on stricter legal requirements of the responsibility of management due to cyber risks or violations for the sake of legal consequences of Article 22, CEO, CIO and CISO owing consequences by Article 23. Avoiding consequences, the following in Table 2.17 must be achieved.

From Table 2.17 and Article 23.4 it can be concluded that:

- Organizations must react swiftly and on time to cyber threat risk incidents that can affect organizations operations effective cybersecurity risk-management and forward reports to the EU Member state national authority,
- Organizations must proactively demonstrate cybersecurity risk-management measures of BCM (Business Continuity Management), supply chain cybersecurity, reporting to the EU Member State national authority, assessment of the development status of cybersecurity risk-management measures, remedial measures, employee cybersecurity awareness training, and others,
- Organizations must set in place a monitoring officer with responsibility for compliance with obligations,
- Organizations must implement acts that are adopted in accordance with the examination procedure referred to in NIS2 Article 39 Committee procedure, paragraph 2.

Table 2.17 Reporting obligations in NIS2 Article 23.4

Reporting Time Line	Obligations
Immediately within 24 h	Upon knowledge of a significant cyber threat risk incident, early warning is indispensable indicating whether there is suspicion that the significant cyber threat risk incident is the result of unlawful or malicious acts or could have cross-border implications
Immediately within 72 h	After becoming aware of a significant cyber threat risk incident, a report is essential, updating information and providing an assessment of the significant cyber threat risk incident, including severity and potential impact, and, if applicable, IoC (indicators of compromise)
At the latest within 1 month	Detailed description of the cyber threat risk incident, including severity and potential impact: (i) Information on the nature of the cyber threat risk incident or underlying cause that likely triggered the cyber threat risk incident (ii) Information on the remedial actions taken and ongoing (iii) If applicable, the potential cross-border impact of the cyber threat risk incident

As a result of the paragraphs in Article 23, each EU Member State shall designate or establish one or more Computer Security Incident Response Teams (CSIRTs) to contribute to the development of confidence and trust and to pro-mote swift and effective operational cooperation among EU Member States. A network of national CSIRTs is established, as described in NIS2 Article 15.1. The CSIRTs may be designated or established within the competent authority. A CSRIT, is a group of IT professionals that provides an organization with services and support in assessment, management and prevention of cybersecurity-related emergencies, as well as coordination of cyber threat risk incident response efforts, to regain control and minimize potential damage. This involves using standardized cyber threat risk incident response cycle activities like the National Institute of Standards and Technology's (NIST) Cybersecurity Framework (CSF) [57] that outline:

- Cybersecurity and cyber threat risk incident handling best practices,
- Definitions of cyber threat risk incident response and other terms and concepts,
- Requirements and best practices for developing cyber threat risk incident response policies, plans, and procedures,
- Cyber threat risk incident response team models, roles, and responsibilities,
- Information sharing recommendations regarding the media, law enforcement, and other organizations,
- Cyber threat risk incident scenarios for TTX (Tabletop Exercises).

Hence, CSIRT serve as the first responder to cyber threat risk incidents within an organization, and perform vital tasks in identifying, mitigating, reviewing, documenting, and reporting findings to management.

Some large organizations may maintain an in-house, dedicated CSIRT team. But many organizations like SMEs (Small and Medium Enterprise's) cannot sustain the cost of a full SOC (Security Operations Center), so they outsource the CSIRT to external service providers. In smaller organizations, hybrid teams are often assembled to respond to cyber threat risk incidents, where only some of the members are dedicated security staff and others work in IT or other departments.

Another important measure in Article 23 is the IoC (Indicator of Compromise), a piece of digital forensic that suggests a system may have been infiltrated by a cyber threat risk incident. IoC provide CSIRTs with crucial knowledge after there has been a data breach or another breach in cybersecurity. Examples of IoCs are:

- Network traffic anomalies:
- Unusual sign-in attempts,
- Privilege account irregularities,
- Changes to system configurations,
- Unexpected software installations or updates,
- Numerous requests for the same file,
- Unusual Domain Name Systems requests,
- And others.

2.5.5 Article 24 Use of European Cybersecurity Certification Schemes

Article 24 Use of European Cybersecurity Certification Schemes Describes this topic in three paragraphs self-explanatory.

1. In order to demonstrate compliance with particular requirements of Article 21, Member States may require essential and important entities to use particular ICT products, ICT services and ICT processes, developed by the essential or important entity or procured from third parties, that are certified under European cybersecurity certification schemes adopted pursuant to Article 49 of Regulation (EU) 2019/881.

Furthermore, Member States shall encourage essential and important entities to use qualified trust services.

2. The Commission is empowered to adopt delegated acts, in accordance with Article 38, to supplement this Directive by specifying which categories of essential and important entities are to be required to use certain certified ICT products, ICT services and ICT processes or obtain a certificate under a European cybersecurity certification scheme adopted pursuant to Article 49 of Regulation (EU) 2019/881. Those delegated acts shall be adopted where insufficient levels of cybersecurity have been identified and shall include an implementation period.

Before adopting such delegated acts, the Commission shall carry out an impact assessment and shall carry out consultations in accordance with Article 56 of Regulation (EU) 2019/881.

3. Where no appropriate European cybersecurity certification scheme for the purposes of paragraph 2 of this Article is available, the Commission may, after consulting the Cooperation Group and the European Cybersecurity Certification Group, request ENISA to prepare a candidate scheme pursuant to Article 48(2) of Regulation (EU) 2019/881.

2.5.6 Article 25 Standardization

Article 25 Standardization Describes this topic in two paragraphs and include a detailed separate subsequently reflection to practical solutions.

1. In order to promote the convergent implementation of Article 21(1) and (2), Member States shall, without imposing or discriminating in favor of the use of a particular type of technology, encourage the use of EU-European and international standards and technical specifications relevant to the security of network and information systems.
2. ENISA, in cooperation with Member States, and, where appropriate, after consulting relevant stakeholders, shall draw up advice and guidelines regarding the technical areas to be considered in relation to paragraph 1 as well as regarding already existing standards, including national standards, which would allow for those areas to be covered.

Table 2.18 Examples of international cybersecurity standards

Cybersecurity Directive Bodies	Specific Area of Directive
IEC: International Electrotechnical Commission	IEC 62443: Automating Cybersecurity and Compliance for International Standards and Technical Specifications (ICAS), the cybersecurity standard for design of Industrial Control Systems
ISO: International Organization for Standardization	ISO establish best practices in relation to the implementation, maintenance and *management* of ISMS (Information *Security Management System*)
NIST CSF: National Institute of Standards and Technology Cybersecurity Framework	Consists of standards, guidelines and best practices to enable management to minimize cybersecurity risks. NIST CSF is designed to be integrated into existing cybersecurity processes in organizations and industries
MITRE ATT&CK (Adversarial Tactics, Techniques & Common Knowledge)	Curated knowledge base to track tactics & techniques used by cybercriminals across entire attack lifecycle. Framework is more than data collection: a tool for strengthening organization's security posture
ISO/IEC	ISO/IEC 27001: Standard include development requirements for ISMS to implement cybersecurity controls and conducting risk assessments

Protentional international standards and technical specifications relevant to cybersecurity of network and information systems, are listed in Table 2.18.

2.6 Chapter V Jurisdiction and Registration

Article 26 Jurisdiction and Territoriality Is expressed in 5 paragraphs. The following only refers to paragraph 1 with its subparagraphs:

1. Entities falling within the scope of this Directive shall be considered to fall under the jurisdiction of the Member State in which they are established, except in the case of:

    (a) providers of public electronic communications networks or providers of publicly available electronic communications services, which shall be considered to fall under the jurisdiction of the Member State in which they provide their services;

    (b) DNS service providers, TLD name registries, entities providing domain name registration services, cloud computing service providers, data centre service providers, content delivery network providers, managed service providers, managed security service providers, as well as providers of online marketplaces, of online search engines or of social networking services platforms, which shall be considered to fall under the jurisdiction of the Member State in which they have their main establishment in the Union under paragraph 2;

    (c) public administration entities, which shall be considered to fall under the jurisdiction of the Member State which established them.

Article 27 Registry of Entities Is expressed in 5 paragraphs. The following only refers to paragraph 2 with its subparagraphs:

2. Member States shall require entities referred to in paragraph 1 to submit the following information to the competent authorities by 17 January 2025:

   (a) the name of the entity;
   (b) the relevant sector, subsector and type of entity referred to in Annex I or II, where applicable;
   (c) the address of the entity's main establishment and its other legal establishments in the Union or, if not established in the Union, of its representative designated pursuant to Article 26(3);
   (d) up-to-date contact details, including email addresses and telephone numbers of the entity and, where applicable, its representative designated pursuant to Article 26(3);
   (e) the Member States where the entity provides services;
   (f) the entity's IP ranges.

The Article lay out registry requirements of entities as a sequence of steps, as illustrated in Table 2.19.

Table 2.19 Examples of registry requirements

Initial situation	Measures
Relevance/ Concern	Clarify whether the organization belongs to the sectors listed in NIS2 Annex I and II
	Organizations are not informed whether they are affected by NIS2
	Organizations must independently assess their own NIS2 impact based on specified sectoral and size-related aspects
	Clarify whether customers belong to the sectors listed in in NIS2 Annex I and Annex II
Responsibility	Clarify who from the management is operationally responsible for implementing NIS2
	Organization must implement measures to prevent and mitigate damage in order to reduce any possible impacts (if necessary, bring in external consultants)
Reporting Obligation	Clearly define responsibility:
	Changes to registration data (Article 3): If registration data stored with national competent authorities change, changes must be submitted within 2 weeks
	Significant cyber threat risk incident (Article 23): If a cyber threat risk incident falls into this category, it must be reported to national authorities and, if applicable, customers or users of the company's own services
	Voluntary reports (Article 30): In addition to reporting obligations, there is the option for organizations to voluntarily report cyber threat risk incidents, regardless of whether they are affected by NIS2:
	• Cyber threat risk incidents that are not significant, cyber threat risk incidents and near-misses
	• Organizations not affected by the directive in the event of significant cyber threat risk incidents, cyber threat risk incidents and near-misses
	Vulnerabilities can be reported by individuals/organizations—even anonymously—to the relevant national authorities

2.7 Chapter VI Information Sharing

Article 28 Database of Domain Name Registration Data Is expressed in 6 paragraphs that are self-explanatory. The following only refers to paragraph 1 and 2 with its subparagraphs:

1. For the purpose of contributing to the security, stability and resilience of the DNS, Member States shall require TLD name registries and entities providing domain name registration services to collect and maintain accurate and complete domain name registration data in a dedicated database with due diligence in accordance with Union data protection law as regards data which are personal data.
2. For the purposes of paragraph 1, Member States shall require the database of domain name registration data to contain the necessary information to identify and contact the holders of the domain names and the points of contact administering the domain names under the TLDs. Such information shall include:
   (a) the domain name;
   (b) the date of registration;
   (c) the registrant's name, contact email address and telephone number;
   (d) the contact email address and telephone number of the point of contact administering the domain name in the event that they are different from those of the registrant.

2.7 Chapter VI Information Sharing

Article 29 Cybersecurity Information-Sharing Arrangements Is expressed in 5 paragraphs that are self-explanatory. The following only refers to paragraph 1 with its subparagraphs:

1. Member States shall ensure that entities falling within the scope of this Directive and, where relevant, other entities not falling within the scope of this Directive are able to exchange on a voluntary basis relevant cybersecurity information among themselves, including information relating to cyber threats, near misses, vulnerabilities, techniques and procedures, indicators of compromise, adversarial tactics, threat-actor-specific information, cybersecurity alerts and recommendations regarding configuration of cybersecurity tools to detect cyber-attacks, where such information sharing:
   (a) aims to prevent, detect, respond to or recover from incidents or to mitigate their impact;
   (b) enhances the level of cybersecurity, in particular through raising awareness in relation to cyber threats, limiting or impeding the ability of such threats to spread, supporting a range of defensive capabilities, vulnerability remediation and disclosure, threat detection, containment and prevention techniques, mitigation strategies, or response and recovery between public and private entities.stages or promoting collaborative cyber threat research.

There is a vast range of communication methods that organizations use to share information that range from face-to-face conversations, instant messaging ser-vices, email, video conferencing, phone calls, and even social media. Thus, information sharing is a crucial feature of operative communication in organizations that involves exchanging relevant information between team members, departments, or even organizations. One specific issue of information sharing is sharing critical information about cyber threat risk incidents and vulnerabilities to plan, process, and develop cybersecurity measures. That's because cyber threat risk incidents become ever sophisticated and fundamentally changed the cybersecurity landscape, so that cybersecurity information sharing is a new practice in organizations detecting contemporary cybercriminal attacks and mitigating their impact [58]. Consequently, new paradigms are required detecting contemporary attacks and mitigating their impact, achieved by information sharing. This is a crucial step acquiring a thorough understanding of large-scale cyber-attack scenarios, and is seen as one key concept to protect network and information systems cybersecurity. For this purpose, specific steps are necessary, as reported in [58]:

- Efficient cooperation and coordination of cyber defense. Due to increased network and information system complexity, attack surfaces, and sophistication of cyber threat risk incidents, coordinated cyber defense is mainly realized through information sharing [59, 60],
- Legal and regulatory landscape directives required for information sharing. EU and US recently begun to create a set of directives and regulations,
- Standards and specifications developed must be compliant with legal requirements. NIST, ENISA, ETSI as standard specifically designed for IoT devices, ISO, NIS2, and others, already released documents to start this effort,
- Based on standards and specifications regional and international implementations need to be realized and integrated. Important work contributing to this step has been performed by CERTs (Cybersecurity Emergency Response Team) and national cyber security centers that are the first responders in the event of cyber threat risk incidents so far.
- Integration of sharing protocols and management tools on organizations technical layer need to be selected and set into practice. It is essential that selected technical means are compatible to organizational processes and can be handled appropriately.

Article 30: Voluntary Notification of Relevant Information Is expressed in 2 paragraphs that are self-explanatory.

1. Member States shall ensure that, in addition to the notification obligation provided for in Article 23, notifications can be submitted to the CSIRTs or, where applicable, the competent authorities, on a voluntary basis, by:

(a) essential and important entities with regard to incidents, cyber threats and near misses;
(b) entities other than those referred to in point (a), regardless of whether they fall within the scope of this Directive, with regard to significant incidents, cyber threats and near misses.

2. Member States shall process the notifications referred to in paragraph 1 of this Article in accordance with the procedure laid down in Article 23. Member States may prioritize the processing of mandatory notifications over voluntary notifications.

Where necessary, the CSIRTs and, where applicable, the competent authorities shall provide the single points of contact with the information about notifications received pursuant to this Article, while ensuring the confidentiality and appropriate protection of the information provided by the notifying entity. Without prejudice to the prevention, investigation, detection and prosecution of criminal offences, voluntary reporting shall not result in the imposition of any additional obligations upon the notifying entity to which it would not have been subject had it not submitted the notification.

2.8 Chapter VII Supervision and Enforcement

Article 31 General Aspects Concerning Supervision and Enforcement Is expressed in 5 paragraphs that are self-explanatory. The following only refers to paragraph 1 with its subparagraphs:

1. Entities falling within the scope of this Directive shall be considered to fall under the jurisdiction of the Member State in which they are established, except in the case of:

    (a) providers of public electronic communications networks or providers of publicly available electronic communications services, which shall be considered to fall under the jurisdiction of the Member State in which they provide their services,
    (b) DNS service providers, TLD name registries, entities providing domain name registration services, cloud computing service providers, data center service providers, content delivery network providers, managed service providers, managed security service providers, as well as providers of online marketplaces, of online search engines or of social networking services platforms, which shall be considered to fall under the jurisdiction of the Member State in which they have their main establishment in the Union under paragraph 2,
    (c) public administration entities, which shall be considered to fall under the jurisdiction of the Member State which established them.

Article 32 Supervisory and Enforcement Measures in Relation to Essential Entities Is expressed in 10 paragraphs that are self-explanatory. The following only refers to paragraph 1:

1. Member States shall ensure that the supervisory or enforcement measures imposed on essential entities in respect of the obligations laid down in this Directive are effective, proportionate and dissuasive, taking into account the circumstances of each individual case.

Article 33 Supervisory and Enforcement Measures in Relation to Important Entities Is expressed in 8 paragraphs that are self-explanatory. The following only refers to paragraph 1:

1. When provided with evidence, indication or information that an important entity allegedly does not comply with this Directive, in particular Articles 21 and 23 thereof, Member States shall ensure that the competent authorities act, where necessary, through ex post supervisory measures. Member States shall ensure that those measures are effective, proportionate and dissuasive, considering the circumstances of each individual case.

Article 34 General Conditions for Imposing Administrative Fines on Essential and Important Entities Is expressed in 6 paragraphs that are self-explanatory. The following only refers to paragraph 1:

1. Member States shall ensure that the administrative fines imposed on essential and important entities pursuant to this Article in respect of infringements of this Directive are effective, proportionate and dissuasive, considering the circumstances of each individual case.

In this context, the paragraph refers administrative fines on essential and important entities based on situation-related boundary conditions, as shown in Table 2.20.

Article 35 Infringements Entailing a Personal Data Breach Is expressed in 3 paragraphs that are self-explanatory. The following only refers to paragraph 1:

1. Where the competent authorities become aware in the course of supervision or enforcement that the infringement by an essential or important entity of the obligations laid down in Articles 21 and 23 of this Directive can entail a personal data breach, as defined in Article 4, point (12), of Regulation (EU) 2016/679 which is to be notified pursuant to Article 33 of that Regulation, they shall, without undue delay, inform the supervisory authorities as referred to in Article 55 or 56 of that Regulation.

Here, a cyber threat risk incident can lead to accidental or unlawful destruction, loss, alteration, unauthorized disclosure of, or access to, personal data transmitted, stored or otherwise processed. Regardless of whether or not a cyber threat risk incident needs to be notified to the supervisory authority, the organizations

2.8 Chapter VII Supervision and Enforcement

Table 2.20 Boundary conditions for fines

Activity	Essential Entities	Important Entities
Supervision by Authorities	Proactive supervision, e.g., regular security audits	Reactive supervision following indications of violations, e.g., targeted security checks
Fines for violations	Maximum at least €ten million or 2% of worldwide turnover	Maximum amount of at least € seven million or 1.4% of worldwide turnover
Whom it may concern	Large companies from Annex I > € 249 million turnover and >€43 million balance sheet Special cases independent of size: e.g. DNS service providers, central government, KRITIS, and facilities classified as "essential" by the state	Large companies from Annex I > €249 million turnover and >€43 million balance sheet Medium-sized companies from Annex I or Annex II at least 50 employees, or >€ten million turnover and >€ten million balance sheet not a large company Special cases independent of size: institutions classified as "important" by the state

responsible controller must keep documentating any personal data breaches, comprising the facts relating to the personal data breach, its effects and the remedial action taken. That documentation shall enable the supervisory authority to verify compliance with Article 35. Furthermore, it has to be noted that after an initial notification, the responsible controller of the organization should update the supervisory authority if a follow-up investigation uncovers evidence that the cyber threat risk incident was contained and no other cyber threat risk incident actually occurred. This information could then be added to the information that already exist to the supervisory authority and the cyber threat risk incident recorded accordingly as not being a data breach incident. In this case, there is no penalty for reporting a cyber threat risk incident that ultimately transpires not to be a data breach incident.

Article 36 Penalties Is expressed in 1 paragraph that is self-explanatory:

1. Member States shall lay down rules on penalties applicable to infringements of national measures adopted pursuant to this Directive and shall take all measures necessary to ensure that they are implemented. The penalties provided for shall be effective, proportionate and dissuasive. Member States shall, by 17 January 2025, notify the Commission of those rules and of those measures and shall notify it, without delay of any subsequent amendment affecting them.

Article 37 Mutual Assistance Is expressed in 2 paragraphs that are self-explanatory. The following only refers to paragraph 1 with its subparagraphs:

1. Where an entity provides services in more than one Member State, or provides services in one or more Member States and its network and information systems are located in one or more other Member States, the competent authorities of the

Member States concerned shall cooperate with and assist each other as necessary. That cooperation shall entail, at least, that:

(a) the competent authorities applying supervisory or enforcement measures in a Member State shall, via the single point of contact, inform and consult the competent authorities in the other Member States concerned on the supervisory and enforcement measures taken;
(b) a competent authority may request another competent authority to take supervisory or enforcement measures;
(c) a competent authority shall, upon receipt of a substantiated request from another competent authority, provide the other competent authority with mutual assistance proportionate to its own resources so that the supervisory or enforcement measures can be implemented in an effective, efficient and consistent manner.

Mutual assistance in criminal matters is a process, generally governed by treaty or authorized by domestic law, by which countries seek and provide information that may be used as evidence in criminal cases. It's an important authority competency allows authorities to be more effective by supporting one another. All competent authorities can, and should, engage in mutual support.

2.9 Chapter VIII Delegated and Implementing Acts

Article 38 Exercise of Delegation Is expressed in 6 paragraphs that are self-explanatory. The following only refers to paragraph 6:

6. A delegated act adopted pursuant to Article 24(2) shall enter into force only if no objection has been expressed either by the European Parliament or by the Council within a period of 2 months of notification of that act to the European Parliament and to the Council or if, before the expiry of that period, the European Parliament and the Council have both informed the Commission that they will not object. That period shall be extended by 2 months at the initiative of the European Parliament or of the Council.

Article 39 Committee Procedure Is expressed in 3 paragraphs that are self-explanatory. The following only refers to paragraph 3:

3. Where the opinion of the committee is to be obtained by written procedure, that procedure shall be terminated without result when, within the time-limit for delivery of the opinion, the chair of the committee so decides or a committee member so requests.

2.10 Chapter IX Final Provisions

Article 40 Review Is expressed self-explanatory.

By 17 October 2027 and every 36 months thereafter, the Commission shall review the functioning of this Directive, and report to the European Parliament and to the Council. The report shall in particular assess the relevance of the size of the entities concerned, and the sectors, subsectors and types of entity referred to in Annexes I and II for the functioning of the economy and society in relation to cybersecurity. To that end and with a view to further advancing the strategic and operational cooperation, the Commission shall take into account the reports of the Cooperation Group and the CSIRTs network on the experience gained at a strategic and operational level. The report shall be accompanied, where necessary, by a legislative proposal.

Article 41 Transposition Is expressed in 2 paragraphs that are self-explanatory.

1. By 17 October 2024, Member States shall adopt and publish the measures necessary to comply with this Directive. They shall immediately inform the Commission thereof. They shall apply those measures from 18 October 2024.
2. When Member States adopt the measures referred to in paragraph 1, they shall contain a reference to this Directive or shall be accompanied by such reference on the occasion of their official publication. The methods of making such reference shall be laid down by Member States.

Article 42 Amendment of Regulation (EU) No 910/2014 Is expressed self-explanatory.

In Regulation (EU) No 910/2014, Article 19 is deleted with effect from 18 October 2024.

Article 43 Amendment of Directive (EU) 2018/1972 Is expressed self-explanatory.

In Directive (EU) 2018/1972, Articles 40 and 41 are deleted with effect from 18 October 2024.

Article 44 Repeal Is expressed self-explanatory.

Directive (EU) 2016/1148 is repealed with effect from 18 October 2024. References to the repealed Directive shall be construed as references to this Directive and shall be read in accordance with the correlation table set out in Annex III.

Article 45 Entry into Force Is expressed self-explanatory.

This Directive shall enter into force on the twentieth day following that of its publication in the Official Journal of the European Union.

Article 46 Addresses Is expressed self-explanatory.

This Directive is addressed to the Member States.

2.11 Exercises

- What is meant by the term *Network and Information Security?*
    - Describe the characteristics, capabilities and requirements of Network and Information Security in the context of NIS2
- What is meant by the term *Cybersecurity*?
    - Describe the characteristics and capabilities of Cybersecurity
- What is meant by the term *Near Miss*?
    - Describe the characteristics and capabilities of Near Miss
- What is meant by the term *Cyber Threat Risk Incident*?
    - Describe the characteristics and capabilities of cyber Threat Risk Incidents
- What is meant by the term *Vulnerability*?
    - Describe the characteristics and capabilities of Vulnerabilities
- What is meant by the term Cybersecurity *Standard*?
    - Describe the characteristics and capabilities of the most important Cybersecurity Standards
- What is meant by the term *Domain Name System (DNS)*?
    - Describe the characteristics and capabilities of DNS
- What is meant by the term *Top Level Domain Registry (TLD)*?
    - Describe the characteristics and capabilities of TLDs
- What is meant by the term *ENISA*?
    - Describe the characteristics and responsibilities of ENISA
- What is meant by the term *CSIRT*?
    - Describe the characteristics and responsibilities of CSIRTs
- What is meant by *Policies on Risk Analysis and Information System Security*?
    - Describe the characteristics and capabilities of Policies on Risk Analysis and Information System Security
- What is meant by the term *Incident Response Handling?*
    - Describe the characteristics and capabilities of Incident Response Handling
- What is meant by the term *Business Continuity (BC)*?
    - Describe the characteristics, capabilities and actions required in Business Continuity

- What is meant by the term *Supply Chain Security*?
  - Describe the characteristics, capabilities and actions required in Supply Chain Security
- What is meant by *Cybersecurity Risk-Management Measures*?
  - Describe the characteristics, capabilities and possible solutions of Cybersecurity Risk-Management Measures
- What is meant by the term *Cyber-Hygiene*?
  - Describe the characteristics, capabilities and actions required in Cyber-Hygiene
- What is meant by the term *Cybersecurity Awareness Training*?
  - Describe the characteristics, capabilities and actions required in Cybersecurity Awareness Training
- What is meant by the term *Cryptography*?
  - Describe the characteristics, capabilities and solutions in Cryptography
- What is meant by the term *Encryption*?
  - Describe the characteristics and capabilities of Encryption
- What is meant by the term *Human Resources Security*?
  - Describe the characteristics, capabilities and solution of Human Resources Security
- What is meant by the term *Multi-Factor Authentication (MFA)*?
  - Describe the characteristics and capabilities of MFA
- What is meant by the term *Secured Communications*?
  - Describe the characteristics and capabilities of Voice, Video and Text Communications
- What is meant by the term *Fines*?
  - Describe the characteristics with regard to Organizations Size, Revenues, and Employees

References

1. Network and Information Security NIS2 – Text Published on the Official Journal of the European Union on 27.12.2022. 2023
2. Press, L.: EU Cybersecurity Directive NIS2 – The Essential Reference. 2024
3. Calder, A.: NIST Cybersecurity Framework: A Pocket Guide. itgp Publ. 2018
4. Cardwell, A.: Navigating NIS2: From Directive to Practice in European Cybersecurity. 2023
5. https://www.quest.com/learn/what-is-cyber-resilience.aspx (accessed 01.2024)
6. NIS2 Directive, 2023. https://eur-lex.europa.eu/legal-content/EN/TXT/PDF/?uri=CELEX:32022L2555 (accessed 01.2023)
7. Albrecht, D.: How to identify Vulnerabilities. 2024. https://fieldeffect.com/blog/how-to-identify-cybersecurity-vulnerabilities (accessed 10.2024)

8. Yasar, K.: Information Security Management System. TechTarget, 2022. https://www.techtarget.com/whatis/definition/information-security-management-system-ISMS (accessed 03.2022)
9. The NIST Cybersecurity Framework (CSF) 2.0, 2024. https://nvlpubs.nist.gov/nistpubs/CSWP/NIST.CSWP.29.pdf (accessed 04.2024)
10. Grance, T., Hash, J., Stevebs, M., O'Neal, K., Bartol, N.: Guide to Information Technology Security Services. NIST Special Publication 800-35, 2003. https://nvlpubs.nist.gov/nistpubs/Legacy/SP/nistspecialpublication800-35.pdf. (accessed 02.2023)
11. 7 Phase of Incident Response: Essential Steps for a Comprehensive Response Plan. https://www.titanfile.com/blog/phases-of-incident-response/. (accessed 09.2023)
12. What is Disaster Recovery. IBM, 2024. https://www.ibm.com/topics/disaster-recovery. (accessed 05.2024)
13. Möller, D. P. F.: Cybersecurity in Digital Transformation – Trends, Methods, Technologies, Applications and Best Practices. Springer Nature, 2023
14. Rock, T.: 11 Key Components of Business Continuity Management (BCM). 20224. https://invenioit.com/continuity/bcm-business-continuity-management/. (accessed 05.2024)
15. Supply Chain Risks, Threats, and Management Strategies. 2024. https://www.bluevoyant.com/knowledge-center/supply-chain-risks-threats-and-management-strategies. (accessed 01.2024)
16. MITRE ATT C&K. https://attack.mitre.org. (accessed 02.2022)
17. NIST Special Publication 800-53 Revision 5 Security and Privacy Controls for Information Systems and Organizations. 2020. https://nvlpubs.nist.gov/nistpubs/SpecialPublications/NIST.SP.800-53r5.pdf. (accessed 02.2022)
18. NIST Special Publication 800-161 Revision 1 Cybersecurity Supply Chain-risk Management Practices for Systems and Organizations. 2022. https://csrc.nist.gov/pubs/sp/800/161/r1/final. (accessed 02.2022)
19. Jüttner, U., Peck, H., Christopher, M.: Supply Chain Risk Management: Outlining an Agenda for Future Research. International Journal of Logistics: Research and Applications, Vol. 6 No. 4, pp. 197-210, 2003
20. Supply Chain Compromise. MITRE ID: T1195 Sub-techniques: T1195.001, T1195.002, T1195.003. 2024. https://attack.mitre.org/techniques/T1195/. (accessed 02.2022)
21. Security Notification – USB Removable Media Provided with Conext Combox and Conext Battery Monitor. Schneider Electric 2018
22. New Investigations into the CCleaner Incident Point to a possible third stage that had Keylogger Capacities. Avast Threat Intelligence, 2018. https://blog.avast.com/new-investigations-inccleaner-incident-point-to-a-possible-third-stage-that-had-keylogger-capacities
23. Behavior Monitoring combined with Machine Learning spoils a massive Dofoil coin mining campaign. Windows Defender Research, 2018. https://www.microsoft.com/en-us/security/blog/2018/03/07/behavior-monitoring-combined-with-machine-learning-spoils-a-massive-dofoil-coin-mining-campaign/. (accessed 12.2022)
24. SK Hack by an Advanced Persistent Threat. Command Five Pty Ltd. 2011. https://www.commandfive.com/papers/C5_APT_SKHack.pdf. (accessed 10.2022)
25. O'Gorman, G., McDonald, G.: The Elderwood Project. 2012. https://web.archive.org/web/20190717233006/http://www.symantec.com/content/en/us/enterprise/media/security_response/whitepapers/the-elderwood-project.pdf. (accessed 02.2023)
26. Krivelevich, D., Gil, O.: Top 10 CI/CD Security Risks, 2024. https://www.cidersecurity.io/top-10-cicd-security-risks/. (accessed 07.2024)
27. OWASP Top Ten Project. OWASP 2018. https://owasp.org/www-project-top-ten/. (accessed 07.2024)
28. Wadhwa, P.: Vulnerability Disclosure: Ensuring Transparency and Security. 2024. https://sprinto.com/blog/vulnerability-disclosure/#. (accessed 07.2024)
29. Bültmann, D.: Penetration Testing vs. Bug Bounty Programs – What Are the Differences? 2024. https://www.oneconsult.com/en/blog/news/penetration-testing-vs-bug-bounty-programs/. (accessed 07.2024)

References

30. International Standard ISO/IEC FDIS 27001 – Information Technology – Information Techniques – Information Security Management Systems- Requirements. https://www.iso.org/obp/ui/en/#iso:std:iso-iec:27001:ed-3:v1:en. (accessed 03.2023)
31. ISo/IEC 27003:2017. https://www.iso.org/obp/ui/en/#iso:std:iso-iec:27003:ed-2:v1:en. accessed 03.2023)
32. ISO/IEC 27004:2016. https://www.iso.org/obp/ui/#iso:std:iso-iec:27004:ed-2:v1:en. (accessed 03.2023)
33. ISO/IEC 27005:2022. https://www.iso.org/obp/ui/#iso:std:iso-iec:27005:ed-4:v1:en. (accessed 03.2023)
34. Network and Information System Security Directive (NIS2), 2022. https://eur-lex.europa.eu/legal-content/EN/TXT/PDF/?uri=CELEX:32022L2555&qid=1707918370993. (accessed 03.2023)
35. National Institute of Standards and Technologies Cybersecurity Framework (NIST CSF). https://www.nist.gov. (accessed 06.2022)
36. ISO 27001 – Annex A.14: System Acquisition, Development and Maintenance. https://www.isms.online/iso-27001/annex-a-14-system-acquisition-development-and-maintenance/. (accessed 03.2023)
37. NIS 2 – 10 Minimum Cybersecurity Risk Management Measures. https://www.goodaccess.com/blog/nis2-10-minimum-cybersecurity-risk-management-measures. (accessed 04.2024)
38. ISO 27001 Clause 9.1 Monitoring, Measurement, Analysis, Evaluation. https://hightable.io/iso-27001-clause-9-1-monitoring-measurement-analysis-evaluation-essential-guide/. (accessed 03.2023)
39. 18 Best practices for effective risk management in cyber security. https://esevel.com/blog/risk-management-in-cyber-security. (accessed 11.2023)
40. Trevors, M., Wallen, C.M.: Cyber-Hygiene: A Baseline Set of Practice. Carnegie Mellon University, 2017. https://insights.sei.cmu.edu/documents/4146/2017_017_001_508771.pdf. (accessed 03.2023)
41. Penetration Testing. www.synposis.com. (accessed 05.2023)
42. A Complete Guide to the Phases of Penetration Testing. https://www.cipher.com. (accessed 05.2023)
43. OSINT Framework. https://osintframework.com/. (accessed 05.2023)
44. Penetration Testing Execution Standard, 2012. http://www.pentest-standard.org/index.php/PTES_Technical_Guidelines. (accessed 05.2023)
45. Beyond Traditional Pentesting: Embracing the New Era of Pentest-as-a-Service (PTaaS). Hackerone, 2025 file:///C:/Users/dmo12/Downloads/beyond-traditional-pentesting.pdf. (accessed 01.2025)
46. Watts, S.: Cyber Hygiene: Introduction and Best Practices. 2024. https://www.splunk.com/en_us/blog/learn/cyber-hygiene.html. (accessed 07.2024)
47. Cryptography Policy Management. http://www.scandiatransplant.org/Documentation/iso27002/physical-and-environmental-security-management. (accessed 07.2023)
48. ISO 27001 Annex A 8.24 Use of Cryptography. https://hightable.io/iso27001-annex-a-8-24-use-of-cryptography/. (accessed 07.2023)
49. Top Three Types of User Authentication. https://www.authgear.com/post/top-three-types-of-user-authentication. (accessed 06.2023)
50. Libeer, L.: A Guide to Access Control, 2024. https://www.lansweeper.com/blog/cybersecurity/a-guide-to-access-control/. (accessed 04.2024)
51. ISO 27001 Annex A.7: Human Resource Security. htts://www.isms.online/iso-27001/annex-a-7-human-resource-security/. (accessed 06.2023)
52. Directive (EU) 2022/2555 of the European Parliament and the of the Council of 14 December 2022 on measures for a high common level of cybersecurity across the Union, amending Regulation (EU) No 910/2014 and Directive (EU)
53. EU Parliamentary Question - E-000709/2023(ASW), 2023. https://www.europarl.europa.eu/doceo/document/E-9-2023-000709-ASW_EN.html. (accessed 09.2023)

54. Flachs, S.: The EU Resilience Act. https://industrialcyber.co/expert/the-eu-cyber-resilience-act/. (accessed 10.2024)
55. Keukniet, G.: Leverage IEC 62443 for EU NIS2 Directive Compliance. DNV, https://www.dnv.com/cybersecurity/cyber-insights/leverage-iec-62443-for-eu-nis2-directive-compliance/. (accessed 10.2024)
56. About ENISA – The EU-European Union Agency for Cybersecurity. https://www.enisa.europa.eu/about-enisa. (accessed 10.2024)
57. NIST SP 800-61 Rev. 2: Computer Security Incident Handling Guide, 2024 https://csrc.nist.gov/pubs/sp/800/61/r2/final. (accessed 07.2024)
58. Skopik, F., Settanni, G., Fiedle, R.: A Problem Shared is a Problem Halved: A Survey on the Dimensions of Collective Cyber Defense through Security Information Sharing. Computers and Security, Vol. 60, pp. 154-176, 2016
59. Gal-Or, E., Ghose, A.: The Economy Incentives for Sharing Security Information. In: Information Systems Research, Vol. 16, No.2, pp. 186-208, 2005 https://doi.org/10.1287/isre.1050.0053. (accessed 02.2023)
60. Gordon, L.A., Loeb, M.P., Lucyshyn, M.: Sharing Information on Computer Systems Security: An Economic Analysis. Journal of Accounting and Public Policy, Vol. 22, No. 6, pp. 461-485, 2003

Chapter 3
Cybersecurity Practices for NIS2 Measures

3.1 Risk-Management and Assessment of Effectiveness of Risk-Management Measures

The intention of the Network and Information Security Directive NIS2 is enhancing the cybersecurity maturity level across the European Union (EU) Member States. For this, a broader range of industry sectors, mandating the implementation of cybersecurity measures in risk-management, and imposing a strict incident reporting. Currently a broad variation in maturity of these issues among EU Member States exist, which this directive seeks to align. However, this requires organizations to take appropriate steps to identify, understand and assess cyber risks to network and information systems, enabling sustainable Business Continuity (BC). This includes effective measures of organizations risk-management to cyber secure network and information systems operation and associated communicating activities. The next step requires analyzing confidence in effectiveness of cybersecurity risk-management measures. These boundary conditions must fit with NIS2 cybersecurity preambles [1], as illustrated in Table 3.1.

Details for preambles 49, 51, 57, 58, 59, 83 and 89 [1] **shown in** Table 3.2.

International cybersecurity frameworks applied for risk-management (Preamble 59) are:

- NIST Cybersecurity Framework 1.1 (NIST CSF 1.1) [3],
- MITRE Adversaries Tactics, Techniques and Common Knowledge Framework MITRE ATTACK (MITRE ATT&CK) [4],
- Critical Security Controls (CSC) [5],
- ISO/IEC 27K [6],
- And others.

Table 3.1 Examples of NIS2 preambles

Preamble	Meaning
49	Strengthening cyber resilience with comprehensive cyber-hygiene policies
50	Raising awareness of cybersecurity and cyber-hygiene
51	Using innovative technologies for better cybersecurity
53	**Cybersecurity in the utility sector**
54	**Ransomware protection**
56	Cybersecurity for Small and Medium-Sized Enterprises (SMEs)
57	Promote active cyber protection in national strategies
58	Focus on high-risk vulnerabilities that can cause costly disruptions
59	Simplifying cybersecurity frameworks and regulatory compliance
70	Large-scale cybersecurity incidents
78/82	Cybersecurity risk-management measures
83	Ensuring cybersecurity of network and information systems
85/90	Cybersecurity in the supply chain
89	Integration of cyber-hygiene and cybersecurity technology
98	Encryption as standard for communication cybersecurity
192	Structured approach to reporting and managing incidents

Table 3.2 Execution requirements of NIS2 preambles

Preamble	Meaning
49	Cyber-hygiene policies provide foundations for protecting network and information systems, infrastructure resources, hardware, software, applications, business assets or end-user data cybersecurity upon which entities rely. Cyber-hygiene policies comprising a common set of practices, including software and hardware updates, password changes, management of new installs, limitation of administrator-level access accounts, and backing-up of data, enabling a proactive framework to preparedness and overall safety and cybersecurity in the event of cyber threat risk incidents. ENISA should monitor and analyze EU Member States' cyber-hygiene policies

(continued)

3.1 Risk-Management and Assessment of Effectiveness of Risk-Management Measures

Table 3.2 (continued)

Preamble	Meaning
51	EU Member States should encourage the use of any innovative technology, including Artificial Intelligence (AI), the use of which could improve the detection and prevention of cyber threat risk incidents, enabling resources to be diverted towards cyber threat risk incidents more effectively. EU Member States should therefore encourage in their national cybersecurity strategy activities in research and development to facilitate the use of such technologies, in particular those relating to automated or semi-automated tools in cybersecurity, and, where relevant, the sharing of data needed for training users of such technology and for improving it. The use of any innovative technology, including AI, should comply with EU data protection law, including the data protection principles of data accuracy, data minimization, fairness and transparency, and data cybersecurity, such as state-of-the-art encryption. The requirements of data protection by design and by default laid down in the EU's General Data Protection Regulation (EU) 2016/679) (GDPR) [2], effective May 25, 2018, for the regulation and increased enforcement of privacy and cybersecurity controls of personally identifiable information in the EU, which should be fully exploited
57	As part of their national cybersecurity strategies, EU Member States should adopt policies on the promotion of active cybersecurity protection as part of a wider defensive strategy. Rather than responding reactively, active cyber threat risk incident protection is the prevention, detection, monitoring, analysis and mitigation of network cybersecurity breaches in an active manner, combined with the use of capabilities deployed within and outside the victim network. This could include EU Member States offering free services or tools to certain entities, including self-service checks, detection tools and takedown services. The ability to rapidly and automatically share and understand cyber threat risk incident information and analysis, cyber risk incident risk incident activity alerts, and response action is critical to enable a unity of effort in successfully preventing, detecting, addressing and blocking incidents against network and information systems. Active cybersecurity protection is based on a defensive strategy that excludes offensive measures
58	Since the exploitation of vulnerabilities in network and information systems may cause significant disruption and harm, swiftly identifying and remedying such vulnerabilities is an important factor in reducing risk. Entities that develop or administer network and information systems should therefore establish appropriate procedures to handle vulnerabilities when they are discovered. Since vulnerabilities are often discovered and disclosed by third parties, the manufacturer or provider of ICT products or ICT services should also put in place the necessary procedures to receive vulnerability information from third parties. In that regard, international standards ISO/IEC 30111:2019 Information technology—Security techniques—Vulnerability handling processes and ISO/IEC 29147: 2022 Medical devices—Information to be supplied by the manufacturer, provide guidance on vulnerability handling and vulnerability disclosure. Strengthening the coordination between reporting natural and legal persons and manufacturers or providers of ICT products or ICT services is particularly important for the purpose of facilitating the voluntary framework of vulnerability disclosure. Coordinated vulnerability disclosure specifies a structured process through which vulnerabilities are reported to the manufacturer or provider of the potentially vulnerable ICT products or ICT services in a manner allowing it to diagnose and remedy the vulnerability before detailed vulnerability information is disclosed to third parties or to the public. Coordinated vulnerability disclosure should also include coordination between the reporting natural or legal person and the manufacturer or provider of the potentially vulnerable ICT products or ICT services as regards the timing of remediation and publication of vulnerabilities

(continued)

Table 3.2 (continued)

Preamble	Meaning
59	The Commission, the European Network and Information Security Agency (ENISA) and the EU Member States should continue to foster alignments with international standards and existing industry best practices in the area of cybersecurity risk-management, for example in the areas of supply chain security assessments, information sharing and vulnerability disclosure
83	High critical (essential) and critical (important) entities should ensure the cybersecurity of the network and information systems which they use in their activities. Those systems are primarily private network and information systems managed by the high critical (essential) and critical (important) entities' internal IT staff or the cybersecurity of which has been outsourced. The cybersecurity risk-management measures and reporting obligations laid down in this Directive should apply to the relevant high critical (essential) and critical (important) entities regardless of whether those entities maintain their network and information systems internally or outsource the maintenance thereof
89	High critical (essential) and critical (important) entities should adopt a wide range of basic cyber-hygiene practices, such as zero-trust principles, software updates, device configuration, network segmentation, identity and access management or user awareness, organize training for their staff and raise awareness concerning cyber threat risk incidents, phishing or social engineering techniques. Furthermore, those entities should evaluate their own cybersecurity capabilities and, where appropriate, pursue the integration of cybersecurity enhancing technologies, such as Artificial Intelligence (AI) or Machine-Learning (ML) systems to enhance their capabilities and the cybersecurity of network and information systems

That describe cybersecurity risk-management requirements for organizations. Risk-management means realizing what digital data and assets require risk analysis. Given that fac, digital data and asset transparency is indispensable, so that cyber risk decisions, cyber risk mitigation measures and the extent of measures taken can be assessed and applied with due care to minimize or eliminate probable cyber risks. In order to develop efficient cybersecurity risk-management, criteria to be taken care about are [7]:

- **Risk Identification and Documentation:** Risk-identification is the essential process to determine potential IT-risks, Cyber-risks, Third-party-risks to organizations to document any risks that could keep an organization from reaching its objective. It's the first step in the risk management process, designed to support organizations to understand and plan for potential cyber risks, because it enables organizations to prepare for potentially harmful cyber risks to minimize their impact before they occur. It involves not only determine potentially or probable cyber threat risk incidents, but also its documentation. Risk-documentation is the systematic recording of all potentially or probable cyber threat risk incidents that can detrimentally impact organizations business, ranging from technology breaches and operational hiccups, to legal entanglements and financial losses. Thus, documentation also serves as evidence of organizations risk-management and cyber threat risk incident assessments.

- **Risk Quantification Analysis and Documentation:** Risk quantification analysis involves assessing and quantifying risks (IT-risks, Cyber-risks, Third-party risks) for alignment to business objectives and regulatory requirements with investments in risk mitigation Hence, it's a knowledge-based statistical approach to understand organizations uncertainty or risk in their business venture with regard to cybersecurity, technology, financial issues and other. Quantifying IT-risks or Cyber-risks or Third-party risks enables comparison to other business risks that helps decision makers balance risks and opportunities. With it, Cybersecurity Risk Quantification (CRQ) supports, for example, cyber-risk management frameworks by prioritizing cyber-risks, facilitating communication to risk owners and executives, and aligning with other risk areas. Cyber risks are uncertain future events that expose impact on organizations operational objectives, for example, unexpected loss in cybersecurity, Business Continuity (BC), and others. To conquer this fact, numerical values applied to determine the probability of specific cyber risks and their potential impact that events may have on organizations business. Key figures in risk quantification analysis expressed by risk frequency and risk severity. The frequency-severity-method is applied to determine the expected number of cyber risks an organization has to face during a period of time and how much the impact of the average cyber risks will take. The method multiplies the average number of cyber risks by the average impact of a cyber risk event. With it, sophisticated models used to determine the likelihood, cyber risk events will have severed impacts. Furthermore, the Loss Event Frequency (LEF) and the Loss Event Magnitude (LEM) are too measures also applied in risk quantification. LEF is the probable frequency within a given timeframe, that a cyber threat risk incident will result in loss. Hence LEF describes the frequency at which a cyber threat risk incident agent acts and, when he act, how much loss an organization might face. LEM is the measurement of the consequences of risk. For example, primary losses include all losses from an as-set (its value, liability, volume, and productivity), and secondary losses that include examining internal and external factors, which cause the loss of assets.
Risk quantification documentation is a description of the analysis process in assessing the risk, an outline of evaluations, and detailed explanations on how conclusions were made.
- **Risk Assessment and Documentation:** Risk assessment is a straightforward, structured and systematic process of identifying risks (IT-risks, Cyber-risks, Third-party risks) and evaluating any associated risks within an organization to implement reasonable control measures, suitable to eliminate, reduce or control. A suitable and sufficient risk assessment be carried out prior to a particular activity in order to eliminate, reduce or suitably control any associated risk involved with, or affected by the activity in question. A risk assessment should be reviewed periodically, proportionate to potentially cyber threat risk incidents, and in any case when either the current risk assessment is no longer valid to the ever-evolving cybersecurity risk landscape, as well as significant changes to the specific risk assessment activity. Therefore, relevant risk assessments should be reviewed following cyber threat risk incidents in order to verify if control mea-

sures and level of evaluated risk where appropriate or require amendment. Risk assessment documentation is a detailed description of the analysis process in assessing the risk including [8] date, time, work area, activity being assessed, risk type, risk control measures, risk ratings, additional controls, new risk ratings (residual), action monitored, and explanations how conclusion were made.

3.1.1 Risk Identification and Documentation

Risk identification and documenting is the process that enables an organization reaching its business objective due to understand and plan for potential risk (IT-risks, Cyber-risks, Third-party-risks) through risk-management and risk assessment [9]. This step requires actions to identify, prioritize and document the important objectives like:

- Access points for each digital entity used and data misuse by unauthorized users,
- Data recovery process and control measures,
- Dependency on service providers, tools, supply chain software providers,
- Important and critical data and assets,
- Internal and external potential cyber threat risk incidents,
- Internal and external vulnerabilities,
- Internal processes and records of administrator rights and activity logs of those granted access,
- And others.

In cyber risk identification application different methods used:

- **SWOT Analysis:** Analyze organization's strengths, weaknesses, opportunities and threats with regard to, e.g., cybersecurity. Understanding, where the organization might be vulnerable, SWOT analyze to discover potential cyber risks and plan accordingly.
    - Strengths Analysis: Record positive attributes of the organization's cybersecurity strategy and others. Questions asked to understand strengths are:

        What are the positive qualities?
        What achievements are made?
        What helps to accomplish goals?
        What resources are available?
        What sets the organization apart from others?
        And others.

    - Weakness Analysis: Captures areas of existing improvement or vulnerabilities within the evaluated subject, for example, cybersecurity strategy. Questions asked to understand weaknesses are:

        What makes it difficult to achieve cybersecurity goals?
        What are the areas for improvement in the cybersecurity strategy?

3.1 Risk-Management and Assessment of Effectiveness of Risk-Management Measures 157

What is lacking in the organization: resources, technology, staff?
What is needed to tackle long-term cybersecurity goals?
And others.

- Opportunities Analysis: List processes by which organizations businesses objectively assess the changes in their sector that could affect their current business operations. Questions ask to understand opportunities analysis are:

Are there new competitors entering the market?
What new features from competitors could erode our competitive advantage?
What new needs of customers can we meet?
What economic trends are beneficial to us?
What are the newest technological breakthroughs?
What disruptive technologies might challenge us?
And others.

- Cyber Risk Analysis: Record external cyber risks that could have a negative effect on organizations subjects, for example, cybersecurity? Questions ask to understand cyber threat risk incidents are:

Have competitors a certain edge over the organization in cybersecurity?
Are their potential new competitors on the horizon with a new cybersecurity innovation plan?
Is market cybersecurity health expected to be based on swiftly implementing the NIS2 Directive?
How does audience, industry or market view the organizations cybersecurity strategy?
What hidden achievements must be accomplished?
And others.

SWOT analysis is applied, when reviewing performance, identify areas for improvement, prioritization, and brainstorming.

- **Root Cause Analysis:** Systematic method to determine a cybersecurity problem's primary cause to develop a way to address it. Main steps to perform a root cause analysis are:

  - Specify the cybersecurity problem,
  - Collect essential data,
  - Determine the causal factors, meaning those that led to the cybersecurity issue,
  - Define which factors are root causes and which are simply symptoms,
  - Identify actions to correct the problem,
  - Find solutions that can stop the problem from happening again,
  - Execute the solution.

- Hence, root cause analysis is a tool that promote organization's continuous improvement if used frequently. It delves deeper to understand why cyber risks succeeded in infiltrating organizations systems due to the dynamic nature of

cyber risk vectors. This complexity requires to employ forensics and behavior analysis, often accommodated in cyber threat intelligence (CTI) methods.
- **Monte Carlo Analysis:** Mathematical modeling technique to determine potential cyber risk's likelihood and impact in cybersecurity measures. Using numbers that fit predetermined cybersecurity criteria, the computer simulates various situations with a different cost-functional and schedule to determine the possibilities of completing a cybersecurity maturity level for a particular cost-functional. Thus it stands as a strategic enabler for cybersecurity teams, offering a systematic and dynamic approach to risk assessment and decision-making with its pivotal role in elevating cybersecurity resilience and defense capabilities.
- **Decision Tree Analysis:** Diagram used to clarify and solve a problem that considers several future possible cyber threat risk incidents and analyzes them at one point in time, enabling to explore the different alternatives one decision can lead to. Each branch on the decision tree represents a possible decision or event while the leaves depict possible outcomes. The decision tree allows to visualize the relationship between various events and consider possible advantages and disadvantages before deciding [9].

3.1.2 Risk Quantification and Documentation

Risk quantification is the practice using numerical values to measure the impact of vulnerabilities on business operations and risks (IT-risks, Cyber-risks, Third-party risks) likelihood. This include identifying cyber risks, by analyzing their impact, and plan minimize their impact, which requires to determine and document:

- Probability of potential cyber threat risk incidents,

    Probability is the likelihood of actually exploiting a potential cyber threat risk incident, considering the cyber threat risk incident type, the capability and motivation of the potential cyber threat risk incident source, and the effectiveness of controls, as shown in Table 3.3.

- Impacts and consequences of potential cyber threat risk incidents that compromise the CIA triad features confidentiality, integrity and availability (see Sect. 1.5) of digital data and assets are considerations, as shown for the CIA triad in Table 3.4 and the definition of potential cyber threat risk incident levels in Table 3.5.

- Which digital data, assets, processes, entities and services require mitigation measures,
- Take the appropriate level of care to implement the scope of cybersecurity measures required to effectively and efficiently mitigate or eliminate potential cyber threat risk incidents,

3.1 Risk-Management and Assessment of Effectiveness of Risk-Management Measures 159

Table 3.3 Probability levels vs. frequency of time per incident

Likelihood	Frequency of cyber threat risk incidents
Most likely	Cyber threat risk incident occurs very frequently, i.e. more frequently than $\geq 10\%$ of the time per incident
Probably	Cyber threat risk incident occurs frequently, i.e. often between >1% and <10% of the time per incident
Possibly Probably	Cyber threat risk incident occurs less frequently, between >0.1% and <1% of the time per incident
Rare/ Unlikely	Cyber threat risk incident occurs rarely, i.e. <0.1% of the time per incident

Table 3.4 Elements of the CIA triad

Core elements	Properties
Confidentiality	Maintain authorized access and disclosure restrictions, privacy and proprietary information protection measures
Integrity	Maintaining the consistency, accuracy and trustworthiness of information and data throughout its lifecycle
Availability	Elementary function of digital entities to successfully prevent data loss

Table 3.5 Definition of risk levels with possible impacts/consequences

Risk level	Impact/consequences
High risk (not acceptable)	Classified cyber threat risk incidents such as phishing to gain access to targeted networks and information systems must be successfully mitigated in real-time
Medium risk	Cyber threat risk incidents that need to be monitored by mitigation measures. Incidents at this risk level leverage malware such as ransomware, infect EthernalBlue vulnerabilities and other
Low risk (acceptable)	Cyber threat risk incidents must be monitored to detect changes that could increase the risk level. Incidents at this risk level include password breaches, insider cyber threat risk incidents, and others

- Which digital data, assets, processes, entities, services at high risk level,
- What assumptions are appropriate for measuring catastrophic, high, medium and low cyber threat risk incident levels as shown in Table 3.6.

Furthermore, risk quantification standards enable organizations to create cyber defense mechanism that protect their digital data, assets, network and information systems from potential malicious cyber threat risk incidents. This enable organizations cybersecurity teams to identify areas of cybersecurity improvements to maintain a cybersecurity posture and adapt to the ever evolving and changing cyber risk landscape. Popular frameworks for risk quantification are NIST SP 800-37 [10] and ISO/IEC 27005 [11] standards.

Table 3.6 Classified cyber risk levels and possible impacts/consequences

Event level	Impact/consequences
Catastrophic	Cyber threat risk incidents classified as severe that have devastating effects and result in loss of trust
Severity	Cyber threat risk incidents classified as serious have serious effects and a loss of trust
Moderate	Cyber threat risk incidents classified as moderate only have a medium effect that can affect trust relationship
Not sever	Cyber threat risk incidents classified as not serious/dangerous only have marginal/minor affect

The NIST SP 800–37 publication [10] describes the Risk-Management Framework (RMF) and provides guidelines for applying the RMF to information systems and organizations for managing cybersecurity and privacy risks. This includes information security categorization like control selection, implementation, and assessment, ongoing information system and common control authorization through the implementation of continuous monitoring processes to provide organizations executives with the necessary information to create efficient risk-management decisions about the systems supporting their missions and business functions into the system development life cycle. In addition, it establishes responsibility and accountability for the controls implemented within organization's information systems and inherited by those systems. The guideline consists of the following process steps [10]:

- **Prepare:** Carry out essential activities at the organization, mission and business process, and information system levels to prepare to manage its cyber risks, applying the respective RMF tasks.
- **Categorize:** Keep risk-management processes and tasks updated by determining the adverse impact to organizations operations and assets, other organizations, and the Nation with respect to the loss of the CIA triad categories (see Sect. 1.5) of organizations systems and the information processed, stored, and transmitted by those systems. Select the right cybersecurity controls to reduce the identified cyber threat risk incidents to an acceptable maturity level, applying the respective RMF tasks.
- **Implement:** Selected controls in the cybersecurity plan for the information system and for the organization and document in a baseline configuration, the specific details of the control implementation, applying the respective RMF tasks. Update the implementation information if any changes occur.
- **Assess:** Determine if the controls selected for implementation are implemented correctly, operating as intended, and producing the desired outcome with respect to meeting the cybersecurity requirements for the information system and the organization reducing cyber threat risk incidents to a desired cybersecurity maturity level, applying the respective RMF tasks.

3.1 Risk-Management and Assessment of Effectiveness of Risk-Management Measures

Likelihood		Very High	High	Medium	Low	Very Low
Business Impact	Very High	8	7	6	5	4
	High	7	6	5	4	3
	Medium	6	5	4	3	2
	Low	5	4	3	2	1
	Very Low	4	3	2	1	0

Fig. 3.1 Likelihood and impact numbers of cyber risks

- **Authorize:** Provide organizational accountability by requiring a management official to determine if the cybersecurity risk incidents, including supply chain cyber-risks, to organizational operations and assets, other organizations, or the Nation based on the operation of a digital system or the use of common controls, is acceptable to a desired cybersecurity maturity level, applying the respective RMF tasks.
- **Monitor:** Maintain an ongoing cybersecurity situational awareness posture of the network and information system and the organization in support of risk-management decisions and keep track of changes, applying the respective RMF tasks.

The ISO/IEC 27005:2022 publication [11] describes the information security, cybersecurity and privacy protection as guidance on managing information cybersecurity risks, to assist organizations to address information cybersecurity risk requirements in ISO/IEC 27001 to conduct risk assessments, and mitigate cyber threat risk incidents. It addresses both qualitative and quantitative cyber threat risk incident analysis methods and recommends using either one or a combination of both, depending on the specific scenario. For this purpose, the guideline uses a cyber risk matrix like the likelihood of a cyber threat risk incident scenario and maps it against the business impact, as illustrated in Fig. 3.1. To calculate a quantitative cyber risk, data from various sources gathered. However, the quality of this method relies heavily on the accuracy of the numerical values.

3.1.3 Risk Assessment and Documentation

Risk assessment is a systematic process of identifying risks (IT-risks, Cyber-risks, Third-party risk) and vulnerabilities and implementing reasonable control measures to mitigate or remove their hazardous business impact. Therefore, the goal of this process is to determine what measures should be implemented to mitigate or remove those potential, for example, cyber risks. Organizations often use a Risk Assessment Framework (RAF) to prioritize and share the details of the assessment, including any cyber threat risk incidents to their IT systems. Form this purpose the RAF contain three important components:

- Shared Vocabulary: Common cross-department vocabulary like a risk language, becoming more commonplace among organization's leadership circle, creating more valuable and meaningful risk-driven conversations, to build a robust RAF,
- Consistent Assessment Methods: Include brainstorming, checklists, interviews, surveys, SWOT analysis, scenario analysis, Root Cause Analysis, Fault Tree Analysis, and Monte Carlo simulation. Since each method has its own advantages and disadvantages, the ones that suit organizations purpose, scope, and resources should be applied,
- Reporting System: Formal document that provides detailed findings from a cybersecurity control assessment, focusing on the determination of cyber risks to organizations system(s) and its operating environment, as well as the probable impact to the organization of those cyber risks occurring or the impact to the organization should they occur. It is produced as a result of a risk assessment process and informs the decision-making process for addressing identified weaknesses or deficiencies in the cybersecurity assessment report.

RAF enables organization to identify cyber risks and any business assets put at cyber risk by these potential cyber threat risk incidents, as well as potential impact if these cyber threat risk incidents come to fruition.

There are several RAF used as industry standards, such as [12]:

- **Factor Analysis of Information Risk (FAIR):** Cyber threat risk incident quantification model, founded to support businesses evaluate information cybersecurity risks [13],
- **Committee of Sponsoring Organizations (COSO):** Establish an Enterprise Risk Management Framework, used for internal controls to be integrated into business processes. In 2023 COSO issued supplemental guidance for organizations to achieve effective internal control over sustainability reporting (ICSR), using the globally recognized COSO Internal Control-Integrated Framework (ICIF) [14].
- **Control Objectives:** For Information and Related Technology (COBIT) from the Information Systems Audit and Control Association (ISACA) created the COBIT framework to bridge the crucial gap between technical issues, business risks and control requirements [15].
- **Operationally Critical Threat, Asset and Vulnerability Evaluation (OCTAVE):** Developed by the Computer Emergency Readiness Team, is a framework for identifying and managing information cybersecurity risks [16]
- **Risk Management Guide**: For Information Technology Systems from the National Institute of Standards (NIST) [3],

- **Threat Assessment and Remediation Analysis (TARA):** Us an engineering methodology used to identify and assess cyber vulnerabilities and select countermeasures effective at mitigating those vulnerabilities. TARA is part of a MITRE portfolio of Systems Security Engineering (SSE) practices that contribute to achievement of Mission Assurance (MA) for systems during the acquisition process [17].

These risk assessment and risk-management frameworks use different approaches to assess risk. For example, an information cybersecurity risk assessment framework will assess IT-risks like vulnerabilities, compliance, financial, operational and strategic risks.

However, risk assessment can vary widely, depending on the cyber threat risk incidents unique to an organizations business and their compliance rules establishing risk-based controls that follow the CIA triad of information cybersecurity. Given that fact, organizations can cover a number of important steps, regardless of their business type, to prioritize and document risk assessment in a corrected manner:

- Identify and prioritize any potential cyber threat risk incidents if they were to occur,
- Determine and prioritize which business assets would be impacted if the cyber threat risk incident came to fruition,
- Decide and prioritize what could be harmed by the cyber threat risk incidents and how to mitigate or eliminate the impact of these incidents on the organizations business assets. Thus, evaluate and prioritize the cyber threat risk incidents and decide to control measures,
- Record and prioritize the risk assessment results and file them as easily accessible official organizations documents that include details on potential cyber threat risk incidents, their associated likelihood of impact and plans to mitigate or prevent the impact,
- Review, prioritize and update the risk assessment regularly since potential cyber threat risk incidents and their controls can change swiftly in business environments. It's important for organizations to update their risk assessments regularly to adapt to these changes.

The risk assessment matrix shows the likelihood of cyber risks and their potential impacts. In the following example, illustrated in Fig. 3.2, likelihood refers to the possibility level that an organization be attacked, by cyber threat risk incidents if exposed to cyber threat risk incidents, while impact refers to the severity of the cyber threat risk incident occurred.

By applying the risk assessment steps mentioned, the organization can manage any potentially cyber threat risk incident to its business. Hence, get prepared with a risk assessment plan while investing time to identify cyber threat risk incidents facing the business and figure out how to manage them.

		Likelihood			
		Extremely likely	Likely	Unlikely	Very unlikely
Impact	Fatal	High	High	High	Medium
	Major Injuries	High	High	Medium	Medium
	Minor injuries	High	Medium	Medium	Low
	Neglectable injuries	Medium	Medium	Low	Low

Fig. 3.2 Likelihood and impact if cyber threat risk incidents came to fruition

3.1.4 Cybersecurity and Data Risk-Management Approach

Main concern in cybersecurity and data risk-management is to protect against cyber risks and vulnerabilities. Therefore, protect digital data, assets, processes, entities services, and others from unauthorized access, use, manipulation, or monetary extortion is the key issue of any cybersecurity and data risk-management approach. However, this is a continuous process, including identifying potentially cyber risks, assessing their likelihood and impact, and implementing measures to defend against or mitigate the identified cyber threat risk incidents. These measures enable to identify, assess, mitigate, and continuously monitor cybersecurity-relevant entities in organizations:

- **Prioritize:** Prioritize and document digital data and assets to be protected and potentially cyber threat risk incidents they are exposed to. This includes assessment of processes, systems, services, networks, applications, and others. Central approach is enabling responsible internal cybersecurity teams of organizations, being cybersecurity situational aware and understand potential cyber threat risk incidents, and prioritize strategic cybersecurity actions and investments to the management.
- **Assess:** After identifying cyber threat risk incidents, internal cybersecurity teams must assess likelihood and their potential impact through vulnerability analysis against cyber threat risk incidents that may be exploited. This support measures to better understand the potential impact of cyber threat risk incidents and enables to prioritize damage mitigation.
- **Monitoring:** Continuous process based on implemented cybersecurity measures. Organizations cybersecurity team continuously monitors processes, systems, services, networks, and others for probable cyber risk. This is to ensure that measures taken to limit damage are effective and also new risk forms are identified and averted in a timely manner. Organizations cybersecurity team supports sustainable cybersecurity and digital data postures to respond effective to new cyber threat risk incidents.

3.1 Risk-Management and Assessment of Effectiveness of Risk-Management Measures 165

- **Damage Limitation:** Organizations cybersecurity teams prioritize measures to limit identified cyber threat risk incidents based on assessments carried out:
  - Implementation of cybersecurity controls,
  - Regular updating of hardware and software resource versions,
  - Regular cybersecurity training of employees and responsible cybersecurity team personnel, and development cyber threat risk incident response and recovery plans.

As a result of the previous steps the following results occur:

- **Improved Cybersecurity:** Cybersecurity and data risk-management enables organizations cybersecurity teams improving cybersecurity posture by detecting potentially cyber threat risk incidents and implementing measures to swiftly mitigate those risks. Thus, organizations cybersecurity teams can prevent cyber threat risk incidents and cybersecure their digital data and assets by implementing effective cybersecurity controls.
- **Cost-effective:** Cybersecurity and data-risk-management enable cost-effectively management of cyber threat risk incidents to support organizations cybersecurity teams prioritizing cybersecurity investments and allocate cybersecurity resources effective and efficient. Thus, organizational cybersecurity teams achieve major cybersecurity gains with less resource expenditure by prioritizing cybersecurity investments accordingly.
- **Support Compliance with Regulatory Requirements:** Enable organizations cyber risk management to protect digital data and assets to proactively exclude reputational loss issues,
- **Business Continuity (BC):** Organizations cybersecurity teams enables ensuring BC identifying cyber threat risk incidents and developing contingency plans to avoid or mitigate their impact. Thus, cybersecurity contingency plans help to support digital data and assets after cyber threat risk incidents occurred.

Given that facts, developing a contingency plan includes essential steps of proactive measures and ways to respond immediately to cyber threat risk incidents to minimize negative impacts on data and assets. To be proactively prepared to cybersecurity threats with potential data and asset loss, and loss in reputation, Recovery Time Objective (RTO) and Recovery Point Objective (RPO) are decisive criteria for contingency planning.

3.1.5 Contingency Planning as Part of Risk-Governance

Contingency planning (CP) means preparing organizations be ready to respond effective and efficient in case of cyber risks. Against this background, Business Continuity Planning (BCP) is a succeeding strategy that proactive ensure that organizations network and information systems continue to act effective and efficient in cyber risks situations. Thus, BCP ensures organizations can continue to operate their essential business activities. In this sense, CP ensures to know what to do and when during a cyber risk, and have the right resources ready to respond swiftly. This means anticipating what probably types of network and information system cyber risks organization might face and knowing how to manage them when they do occur. Therefore, CP enables organizations business to proactive prepare for the unexpected, which requires decisive measures, because cyber risks can occur at any time and de-rail businesses quite easily, which raise important questions:

- How do organizations know if they can handle these situations?
- How do organizations know if their IT infrastructure is able to recover in time and minimize down-time impact?
- How do organizations replicate data for a near-zero loss?

This means Recovery Time Objective (RTO) comes in. RTO is the longest time an application, computer, network, or information system can be offline after an unexpected cyber threat risk incident come to fruition. It defines the maximum time allowed for restoring regular service levels and resuming typical operations after the probable disruption. Against this background, RTO is a key metric for proactively assessing and improving regular business operation due to cyber risk recovery in the context of Business Continuity (BC).

Another key metric that comes in is the Recovery Point Objective (RPO). RPO refers to the age of data recovered from a backup storage after a successful cyber threat risk incident, to resume regular business operations. However, the point in time before a potential cyber threat risk incident happen, is decisive, to successfully recover data and assets, meaning time elapsed since last reliable back-up. In this context, RPO determines what is the minimum back-up frequency, which result in the effectiveness of recovery from a cyber threat risk incident regarding how much data can be lost after a cyber threat risk incident came to fruition, to proceed with regularly business processes. Therefore, organizations cybersecurity team must decide how less far back the organization can perform sufficient recovery without delaying data loss compared to the expected RTO. The measures that affect RPO are:

- Maximum tolerable data loss for cyber threat risk incident that come to fruition,
- Cybersecurity-situational specific measures when dealing with sensitive digital data and assets,
- Recovery speed affect data storage options for data-cybersecurity (physical files, cloud storage, and othetrs),
- Cost of data loss (monetary and reputation wise) and loss of cybersecurity measures,
- Cost of implementing cyber risks recovery solutions,
- Time to Data (TtD) represent the time required to retrieve backup data and provide recovery locations and systems.

Finally, compliance systems affect cybersecurity provisions for data recovery, data loss, and data confidentiality, integrity and availability (CIA triad).

RTO refers to the maximum acceptable time (duration) during which organizations systems, services, and others can be down after a successful cyber threat risk incident. That is why RTO represent the time span between occurrence of cyber threat risk incidents and restoration of regular business operation. For this purpose, RTO is like RPO decisive in a Disaster Recovery Plan (DRP), a structured approach considering instructions to be followed if a cyber threat risk incident came to fruition in order to restore business or service operations. DRP considers information to effectively support affected systems, devices, resources, tools or services for the recovery process by backups and cybersecurity teams responsible for restoring failed systems, devices, resources, and documentation of potential recovery steps. A checklist for a DRP thus includes identifying critical networks and IT systems, setting priorities for RTOs, and describing the steps required to respond to and restart, and/or restore networks and IT-systems. Once, RTOs decided, the organizations cybersecurity team decides which DRP methods best suited for the actual situation, because DRPs can be tailored specifically to a particular environment, such as:

- **Virtualized DRP:** Enables disaster recovery to be implemented more efficiently and easily.
- **Network DRP:** Due to the complexity of networks, the recovery process must be described step by step and in detail.
- **Cloud DRP:** Can range from cloud backup to full replication.
- **DRP for Data Centers:** Focuses on the data center setup and infrastructure.
- And others.

Hence, DRP covers organization's ability to respond to and recover from unexpected cyber threat risk incidents. DRP also enables critical processes, systems, services and other resources to be used again swift after a disaster recovery. Several steps required by DRP:

- Determination of the scope of the necessary measures and activities for recovery,
- Identification of the most serious vulnerabilities and cyber threat risk incidents as well as critical assets,
- Review the history of unforeseen incidents and/or failures and their treatment,
- Determination of the necessary measures for the DRP, to swiftly respond to and resolving outages,
- Deploy the Incident Response Team (IRT), test the DRP, updating if necessary,
- Conducting a DRP audit,

With it, DRP is a sub-area of BCP, where BCP is focused on restoring essential functions according to Business Impact Analysis (BIA), that predicts consequences of a business disruption and gathers information to develop recovery strategies.

Due to the importance of RPO and RTO essential conditions contain:

- RTO Calculation: Result is speed through which the recovery process must be carried out after a cyber threat risk incident,
- Basis of Loss Tolerance: Process to determine and assess potential impacts of critical cyber-security processes due to cyber risks as part of BCP. This can be

understood as exploratory component for uncovering cyber threat risk incidents and vulnerabilities, but also a planning component for developing strategies to limit the damage of cyber threat risk incidents and to quantify their potential, including loss of the current cybersecurity activity. For that reason, the definition of loss tolerance for cyber threat risks incidents is required.

- Loss Tolerance: Determining how much operating time an organization can lose after a cyber threat risk incident before regular cybersecurity operations can be resumed,
- Complete Inventory of Organization's Cybersecurity Environment (RTO procedure): Including all processes, systems, services, cybersecurity-critical applications, and data. RTO cannot be determined without a precise inventory,
- RPO and RTO Goal: Implement a DRP that in practice is an efficient and effective backup strategy, intended to restore the infiltrated entity to its regular operating or functional state,
- Design a Backup System: Must be practical that possible downtime costs are considered together with ongoing costs for implementing the required DRP,
- Time to Data (TtD): Must have a negative impact on both RPO and RTO. The reason is that RPO can only be completed when backup data is available for disaster recovery. RTO can only be carried out in the case of disaster recovery after backup data is available. Therefore, Time-To-Detect (TTD) should be as small as possible when designing RPO and RTO in order to keep the negative impact as low as possible.

Given that facts, RPO and RTO do not necessarily have the same scope to cybersecure specific organizational processes.

- **Goal for IT Systems:** As small RPO as possible, since no data should be lost,
- **Goal for OT Systems:** To restore OT systems to a functional state as swiftly as possible after a fault. However, the term functional does not describe achieving 100% regular operating status before a fault,

Therefore, an acceptable partial state of recovery refers to a situation that an essential functionality restored and RTO shortened. The remaining missing functions not required for essential functionality, are restored after the required partial state of recovery is achieved. In this regard, recovery goals include the decision how long an organization can be offline and how much data loss can it tolerate. This is determined by RPO that set the maximum amount of data an organization can tolerate, measured in the time from the occurrence of the probable cyber threat risk incident to the last valid data backup. In this sense, RTO refers to the maximum time required to restore regular operations after data loss or failure. Finally, the DRP defines how swiftly recovery takes place after a cyber threat risk incident that unexpectedly paralyzes critical operations and makes access to data inaccessible. This enables a fast return to regular and cybersecure operations and avoids damage to the organization's cybersecurity posture.

To Summarize: RTO and RPO are both critical in cyber threat risk incident recovery planning but serve different purposes. While RTO refers to the maximum

3.1 Risk-Management and Assessment of Effectiveness of Risk-Management Measures 169

Table 3.7 RTO and RPO criteria in cyber threat risk incident recovery planning

Criteria	RTO	RPO
Main focus	Time to restore data	Data that must be recovered
Purpose	Minimize downtime and restore regular business operations	Minimize data loss and ensure data integrity
Boundary condition	Time (e.g., minutes, hours, days)	Time (e.g., minutes, hours, days)
Example	If RTO is 4 h, data must be restored within 4 h of a disruption	If RPO is 60 min, data backup must ensure no more than 60 min of data lost
Impact	Business continuity	Data integrity and consistency
Scope	System recovery and business resumption	Data backup and recovery
Tools and strategy	Disaster recovery plans, failover systems	Backup solutions replication, continuous data protection
Recovery technique	System restoring, failover procedures	Data backups, snapshots, data replication

allowable downtime before operations are significantly impacted, RPO refers to the maximum allowable data loss measured in time. Table 3.7 illustrates the differences in detail.

3.1.6 Compliance Management System

A Compliance Management System (CMS) brings together structures, processes, and measures within organizations that serve to ensure compliance so that organizations do not violate legally binding (external) directives and (internal) requirements. Thus, a CMS is a unified, holistic and integrative whole that, based on a cycle, encompasses legal certainty and the reduction of liability risks, as well as resource optimization and the trust of customers and business partners. This cycle is characterized by four fundamental features, which can be exemplified by a chain of actions, based on the established ISO principle of Plan-Do-Check-Act (PDCA), illustrated in Fig. 3.3.

Against this background, the goal of a CMS is to ensure that all levels within organizations, from the board of directors to employees, adhere to directives and guidelines, as even minor violations can cause reputational and financial damage to organizations. To prevent this, CMS typically uses several principles designed to provide evidence for assessing the appropriateness and effectiveness. This inevitable provides a framework of essential basic elements for optimally considering the scope for action within organizations:

- CMS Culture: The basis for CMS within the company, which is promoted by management behavior,
- CMS Goals: Targets achievable through CMS are defined here,
- CMS Risks: List of potential risks, ranked by probability of occurrence,

Fig. 3.3 Illustration of the ISO PDCA principle

P: *Planning a risk management system*
↓
D: *Identifying business risks*
↓
C: *Prioritizing business risks*
↓
A: *Implementing measures.*

- CMS Program: Implementation of policies and measures that ensure successful CMS execution,
- CMS Organization: Definition of responsibilities and authorities within the framework of CMS,
- CMS Communication: CMS obligations are rolled out to employees,
- CMS Monitoring and Improvement: CMS documentation ensures compliance within the company and its continuous improvement.

For the development of CMS, ISO 37301 was published as a certifiable, global standard for CMS, relevant for compliance. The standard is applicable to all types of organizations, regardless of their size, industry, risk exposure, or global presence. ISO 37301 is based on the established ISO principle of the Plan-Do-Check-Act (PDCA) approach.

3.2 Cybersecurity Frameworks and Criteria

In the era of digital transformation organizations are under pressure in order to comply high and sustainable cybersecurity maturity levels within their business processes and services, for sustainable cybersecurity posture. However, the pace of potential cyber risks is swiftly and today cyber risks are even sophisticated and destructive than ever. All it takes is one successful breach to complicate organizations cybersecurity posture, and spoil its trustability with regard to the CIA triad features (see Sect. 1.5). Against this background, cybersecurity is of utmost priority of national and international cybersecurity postures. Therefore, national and international organizations work on cybersecurity directives and frameworks, like:

- **National Institute of Standards and Technology (NIST):** A division of the U.S. Department of Commerce, working with industry and academia to enhance economic cybersecurity by developing the cybersecurity framework NIST CSF [3],
- **MITRE Adversarial Tactics, Techniques, and Common Knowledge— MITRE ATTACK (MITRE ATT & CK):** Initiative started in 2015 with the goal of providing a globally accessible knowledge base of adversary tactics and techniques based on real-world observations to cybersecurity [4],

3.2 Cybersecurity Frameworks and Criteria

- **Center for Internet Security Critical Security Controls (CISCSC):** Initiated in 2008. responding to experienced extreme data losses. It was developed by the SANS Institute, a private U.S. for-profit company [5],
- **International Standardization Organization and Institute of Electrotechnical Engineers (ISO/IEC):** Worldwide network of national standards bodies that Initiate standards designed to ensure that products and services are safe, reliable, and of high quality, and are compatible with each other. In this sense, standards are the deep knowledge of people with expertise in their subject matter, knowing the needs of their organizations [6].
- **Network and Information Security (NIS2):** *EU-wide legislation on a high common level of cybersecurity* [1]. It provides legal measures to boost the overall high level of cybersecurity in the EU Member States. NIS2 was proposed by the European Commission in December 2020, entered into force at the EU level on January 16, 2023, and requires that the obligations of the Directive must be implemented by the EU Member States as national law.

All cybersecurity directives and frameworks take measures to fill the potential cybersecurity gaps and ramp up cybersecurity, to improve a sustainable cybersecurity posture with regard to the cybersecurity maturity level. The initiated measures in the directives and frameworks measures refers to the cybersecurity situational awareness basics, as outlined in Chap. 1. A major concern is a systematic approach to manage organization's network and information systems cybersecurity, including policies, procedures and others, for managing cybersecurity of their data and assets.

3.2.1 NIST Cybersecurity Framework (NIST CSF)

The United States depends on the reliable functioning of its critical infrastructure and digital systems. Cyber risks exploit the increased complexity and connectivity of critical infrastructure and digital systems, placing the Nation's cybersecurity, economy, and public safety and cyber health at risk. Given that fact, the U.S. National Institute of Standards and Technology created a Cybersecurity Framework (CSF), the NIST CSF, a standardized method for measuring capability in cyber defense and protection. It provides guidelines for organizations to proactively manage cybersecurity risks, identify vulnerabilities, respond to cyber incidents effectively to administer and minimize their cybersecurity risks. The Framework has been used widely to reduce cybersecurity risks since its initial publication as NIST CSF 1.0 in 2014 [2]. Due to current and future cybersecurity challenges, the cybersecurity rules in NIST CSF 1.0, which was designed for critical infrastructure organizations, was enhanced to keep up with increased digitization and an evolving cybersecurity landscape as NIST CSF 2.0 cybersecurity framework [18]. NSF CSF 2.0 [19] integrate the result of years of practical application and analysis of cyber threat risk incidents.

Thereby, it does not merely define a set of cybersecurity measures but forms a systemic approach that integrates cybersecurity into organizations strategic and operational processes. Hence, it's a vital tool for organizations of all sizes and sectors, enabling them anticipate and deal with cyber threat risk incidents to manage and reduce their cybersecurity risks. Because of that, NIS 2.0 introduces a new Govern function, emphasizing the critical role of governance in cybersecurity risk management. The new Govern function consists of 31 out of the 106 subcategories of NIST CSF 2.0, demonstrating the importance of managing cyber risk scorrectly to succeed at preventing, managing, and recovering from cyber-attacks. This change is aligned with the growing liability expectations from management and boards of organizations, demonstrated through the new rules that the Securities and Exchange Commission (SEC) adopted earlier regarding cyber risk management, strategy, governance, and incident disclosure by public companies. SEC cyber disclosure rule is putting the onus on organizations to give investors current, consistent and decision-useful information about how they manage their cyber risks. In this regard, NIST CSF 2.0 requires a comprehensive approach that goes beyond narrowly specialized technical measures. Key updates in NIST CSF 2.0 are:

- Expansion of the governance function,
- In-depth recommendations for supply-chain management,
- Integration with emerging technologies,
- New metrics for evaluating cybersecurity maturity.

Furthermore, NIST CSF 2 includes the following components [18]:

- **CSF Core**: A taxonomy of high-level cybersecurity outcomes that can support any organization manage its cyber risks. CSF core components are a hierarchy of Functions, Categories, and Subcategories that detail each outcome. These outcomes can be understood by a broad audience, are sector-, country-, and technology-neutral, and provide an organization with the flexibility needed to address its unique risks, technologies, and cybersecurity mission considerations.
- **CSF Organizational Profiles**: A mechanism for describing organization's current and/or target cybersecurity posture in terms of the CSF Core's outcomes.
- **CSF Tiers**: Applied to CSF Organizational Profiles to characterize the rigor of an organization's cyber risk governance and management practices. Tiers can also provide context for how an organization views cyber risks and the processes in place to manage those risks.

Against this background, cybersecurity risk-management is the ongoing process of identifying, assessing, and responding to cyber risks. To manage cyber risks, organizations need to understand the likelihood that cyber risks that occur and the potential resulting impacts. With this information, organizations can determine the acceptable level of cyber risks for achieving their organizational objectives and can express this as their cyber risk tolerance.

NIST CSF 2.0 Core Functions

The six NIST CSF 2.0 Core Functions described below, should be performed concurrently and continuously, to form an operational culture that addresses the dynamic cyber risk [18]:

- **Govern (GV)**: The organization's cybersecurity risk management strategy, expectations, and policy are established, communicated, and monitored. GV provides outcomes to inform what an organization should do to achieve and prioritize the outcomes of the other five Core Functions in the context of its requirements and expectations. Governance activities are critical for incorporating cybersecurity into an organization's broader Enterprise Risk Management (ERM) strategy. GV addresses understanding of organizational context like the establishment of cybersecurity strategy and cybersecurity supply chain risk management, the roles, responsibilities, and authorities, the policy, and the oversight of cybersecurity strategy.
- **Identify (ID)**: Organization's current cyber risks are understood. Understanding organization's assets, suppliers, and related cyber risks enables an organization to prioritize its efforts consistent with its risk management strategy and the mission needs identified under GV. This Function also includes the identification of improvement opportunities for the organization's policies, plans, processes, procedures and practices that support cybersecurity risk management to inform efforts under all six functions.
- **Protect (PR)**: Safeguards to manage the organization's cyber risks are used. Once assets and cyber risks are identified and prioritized, PR supports the ability to secure those assets to prevent or lower the likelihood and impact of adverse cybersecurity events, as well as to increase the likelihood and impact of taking advantage of opportunities. Outcomes covered by this function include identity management, authentication, and access control; awareness and training; data security; platform security (i.e., securing the hardware, software, and services of physical and virtual platforms); and the resilience of technology infrastructure.
- **Detect (DE)**: Possible cyber risks and compromises are found and analyzed. DE enables the timely discovery and analysis of anomalies, Indicators of Compromise (IoC), and other potentially cybercriminal incidents that may indicate that cyber threat risk incidents are occurring. This function supports successful cyber threat risk incident response and recovery activities.
- **Respond (RS)**: Actions regarding a detected cyber threat risk incident are taken. RS supports the ability to contain the effects of cyber threat risk incidents. Outcomes within this function cover incident management, analysis, mitigation, reporting, and communication.
- **Recover (RC)**: Assets and operations affected by a cyber threat risk incident are restored. RC supports the timely restoration of normal operations to reduce the effects of cybersecurity incidents and enable appropriate communication during recovery efforts.

The NIST CSF 2.0 Core Functions, their categories and their category identifiers are illustrated in Table 3.8.

Categories in each Core Function describes activities of each functions to achieve their objectives as illustrated in Tables 3.9, 3.10, 3.11, 3.12, 3.13 and 3.14.

The NIST CSF 2.0 framework is result-oriented, but it does not dictate how an organization must achieve its results, but rather enables risk-based implementations. From Tables 3.9, 3.10, 3.11, 3.12, 3.13 and 3.14 it can be seen that the focus of categories and subcategories is always on specific activities, aiming on optimal business outcomes.

NIST CSF 2.0 Profiles

Profiles describes an organization's current and/or target cybersecurity posture in terms of Core outcomes. Organizational Profiles used to understand, tailor, assess, prioritize, and communicate Core outcomes by considering an organization's goal specific objectives, stakeholder expectations, cyber threat risk incident landscape, and requirements. An organization can then prioritize its actions to achieve specific outcomes and communicate that information to stakeholders.

Table 3.8 NIST CSF 2.0 Core Functions, Category names and Identifiers

Function	Category	Identifier
Govern (GV)	Organizational Context	GV.OC
	Risk Management Strategy	GV.RM
	Roles, Responsibilities, and Authorities	G.RR
	Policies	GV.PO
	Oversight	GV.OV
	Cybersecurity Supply Chain Risk Management	GV.SV
Identify (ID)	Asset Management	ID.AM
	Risk Assessment	ID.RM
	Improvement	ID.IM
Protect (PR)	Identity Management, Authentication, and Access Control	PR.AA
	Awareness and Training	PR.AT
	Data Security PR.DS	PR.DS
	Platform Security PR.PS	PR.PS
	Technology Infrastructure Resilience	PR.IS
Detect (DE)	Continuous Monitoring	DE.CM
	Adverse Event Analysis	DE.AE
Respond (RS)	Incident Management	RS.MA
	Incident Analysis	RS.AN
	Incident Response Reporting and Communication	RS.CO
	Incident Mitigation	RS.MI
Recover (RC)	Incident Recovery Plan Execution	RC.RP
	Incident Recovery Communication	RC.CO

3.2 Cybersecurity Frameworks and Criteria 175

Table 3.9 Govern (GC) Categories and number of respective Sub-Identifiers

Function	Categories and numbers of Sub-Identifiers
Govern (GV)	Organizational Context (GV.OC): Circumstances, requirements, expectations, dependencies, legal, regulatory, and contractual requirements surrounding organization's cybersecurity risk management decisions, described in 5 sub-identifiers
	Risk Management Strategy (GV.RM): Organization's priorities, constraints, risk and risk tolerance assumptions are established, communicated, and used to support operational risk decision, described in 7 sub-identifiers
	Roles, Responsibilities, and Authorities (GV.RR): Cybersecurity roles, responsibilities, and authorities to foster accountability, performance assessment, and continuous improvement are established and communicated, described in 4 sub-identifiers
	Policy (GV.PO): Organizational cybersecurity policy is established, communicated, and enforced, described in 2 sub-identifiers
	Oversight (GV.OV): Results of organization-wide cybersecurity risk management activities and performance used to inform, improve, and adjust the risk management strategy, described in 3 sub-identifiers
	Cybersecurity Supply Chain Risk Management (GV.SC): Cyber supply chain risk management processes are identified, established, managed, monitored, and improved by organizational stakeholders, described in 10 sub-identifiers

Table 3.10 Identify (ID) Categories and number of respective Sub-Identifiers

Function	Categories and numbers of Sub-Identifiers
Identify (ID)	Asset Management (ID.AM): Assets that enable organization to achieve business purposes are identified and managed consistent with their relative importance to organizational objectives and the organization's risk strategy, described in 8 sub-identifiers
	Risk Assessment (ID.RA): Cybersecurity risk to organization, assets, and individuals is understood, described in 10 sub-identifiers
	Improvement (ID.IM): Improvements to organizational cybersecurity risk management processes, procedures and activities are identified across all NIST CSF 2.0 Functions by the organization, described in 4 sub-identifiers

Organizational Profile includes one or both of the following [18]:

- **Current Profile:** Specifies Core outcomes that an organization is currently achieving (or attempting to achieve) and characterizes how or to what extent each outcome is being achieved.
- **Target Profile:** Specifies the desired outcomes that an organization has selected and prioritized for achieving its cybersecurity risk-management objectives. Thus, a Target Profile considers anticipated changes to organization's cybersecurity posture, such as new requirements, new technology adoption, and Threat Intelligence (TI) trends.

Table 3.11 Protect (PR) Categories and number of respective Sub-Identifiers

Function	Categories and numbers of Sub-Identifiers
Protect (PR)	Identity Management, Authentication, and Access Control (PR.AA): Access to physical and logical assets limited to authorized users, services, and hardware and managed commensurate with the assessed risk of unauthorized access, described in 6 sub-identifiers
	Awareness and Training (PR.AT): Organization's personnel are provided with cybersecurity awareness and training so that they can perform their cybersecurity-related tasks, described in 2 sub-identifiers
	Data Security (PR.DS): Data are managed consistent with the organization's risk strategy to protect the confidentiality, integrity, and availability of information, described in 4 sub-identifiers
	Platform Security (PR.PS): Hardware, software (e.g., firmware, operating systems, applications), and services of physical and virtual platforms managed consistent with organization's risk strategy to protect CIA measures (see Sect. 1.5), described in 6 sub-identifiers
	Technology Infrastructure Resilience (PR.IR): Security architectures are managed with organization's risk strategy to protect asset CIA measures and organizational resilience, described in 4 sub-identifiers

Table 3.12 Detect (DE) Categories and number of respective Sub-Identifiers

Function	Categories and numbers of Sub-Identifiers
Detect (DE)	Continuous Monitoring (DE.CM): Assets are monitored to find anomalies, Indicators of Compromise (IoC), and other potentially adverse events, described in 9 sub-identifiers
	Adverse Event Analysis (DE.AE): Anomalies, IoC, and other potentially adverse events are analyzed to characterize the events and detect cybersecurity incidents, described in 8 sub-identifiers

Table 3.13 Respond (RS) Categories and number of respective Sub-Identifiers

Function	Categories and numbers of Sub-Identifiers
Respond (RS)	Incident Management (RS.MA): Responses to detected cybersecurity incidents are managed, described in 5 sub-identifiers
	Incident Analysis (RS.AN): Investigations are conducted to ensure effective response and support forensic and recovery activities, described in 4 sub-identifiers
	Incident Response Reporting and Communication (RS.CO): Response activities are coordinated with internal and external stakeholders as required by laws, regulations, or policies, described in 2 sub-identifiers
	Incident Mitigation (RS.MI): Activities are performed to prevent expansion of an event and mitigate its effects, described in 2 sub-identifiers

Table 3.14 Recover (RC) Categories and number of respective Sub-Identifiers

Function	Categories and numbers of Sub-Identifiers
Recover (RC)	Incident Recovery Plan Execution (RC.RP): Restoration activities are performed to ensure operational availability of systems and services affected by cybersecurity incidents, described in 6 sub-identifiers
	Incident Recovery Communication (RC.CO): Restoration activities are coordinated with internal and external parties, described in 2 sub-identifiers

3.2 Cybersecurity Frameworks and Criteria

Therefore, an organization could apply the following steps creating and using Organizational Profiles to continuously support cybersecurity improvement [18].

- **Scope the Organizational Profile:** Document the high-level facts and assumptions on which the Profile will be based to define its scope. An organization can have as many Organizational Profiles as desired, each with a different scope. For example, a Profile could address an entire organization or be scoped to an organization's financial systems or to countering ransomware threats and handling ransomware incidents involving those financial systems.
- **Gather information needed to prepare the Organizational Profile:** Examples of information may include organizational policies, risk-management priorities and resources, enterprise cyber risk profiles, Business Impact Analysis (BIA) registers, cybersecurity requirements and standards followed by the organization, practices and tools (e.g., procedures and safeguards), and work roles.
- **Create the Organizational Profile:** Determine what types of information the Profile should include for the selected CSF outcomes, and document the needed information. Consider the risk implications of the *Current Profile* to inform *Target Profile* planning and prioritization. Also, consider using a Community Profile as the basis for the *Target Profile*.
  - **Analyze gaps between Current and Target Profiles, and create an action plan:** Conduct a GAP-analysis to identify and analyze the differences between the Current and Target Profiles, and develop a prioritized action plan (e.g., risk register, risk detail report, Plan of Action and Milestones (POA&M) to address those gaps. A GAP-analysis is a method of assessing the performance of a business unit to determine whether business requirements or objectives are being met and, if not, what steps should be taken to meet them, e.g., GAP-analysis compliance with legal requirements like NIST CSF 2.0, ISO 27 K, NIS2, and others
- **Implement the action plan, and update the Organizational Profile:** Follow the action plan to address the gaps and move the organization toward the Target Profile. An action plan may have an overall deadline or be ongoing.

Given the importance of continual improvement, an organization can repeat these steps as often as needed.

There are additional uses for Organizational Profiles. For example, a Current Profile can be used to document and communicate the organization's cybersecurity capabilities and known opportunities for improvement with external stakeholders, such as business partners or prospective customers. Also, a Target Profile can help express the organization's cybersecurity risk-management requirements and expectations to suppliers, partners, and other third parties as a target for those parties to achieve.

NIST CSF 2.0 Tiers

Tiers characterize the rigor of organization's cybersecurity risk governance and management practices, and describe how an organization views cybers risks and the processes in place to manage them. Tiers include four steps: Partial (Tier 1), Risk

Informed (Tier 2), Repeatable (Tier 3), and Adaptive (Tier 4). With it, Tiers describe a progression from informal, ad hoc responses to approaches that are agile, risk-informed, and continuously improving. Selecting Tiers enables how an organization will manage its cybersecurity risks. However, Tiers should complement an organization's cybersecurity risk management methodology rather than replace it [18].

Tables 3.15, 3.16, 3.17 and 3.18 contains a notional illustration of the CSF Tiers of an organization's cybersecurity risk governance practices Govern, and cybersecurity risk management practices Identify, Protect, Detect, Respond, and Recover [18].

Progression to higher Tiers is encouraged when cyber risks or mandates are greater or when a Cost-Benefit Analysis (CBA) indicates a feasible and cost-effective reduction of negative cyber risks. A CBA is a process of comparing the projected costs and benefits of a decision to determine its feasibility. Therefore, businesses can determine whether a decision is worthwhile by summing up the

Table 3.15 Illustration of NIST CSF Tier 1

Tier	Cybersecurity risk government	Cybersecurity risk-management
Tier 1: Partial	Application of organizational cybersecurity risk strategy is managed in an ad hoc manner. Prioritization is ad hoc and not formally based on objectives or threat environment	Limited awareness of cybersecurity risks at organizational level Organization implement cybersecurity risk management on an irregular, case-by-case basis Organization may not have processes that enable cybersecurity information to be shared within the organization Organization generally unaware of cybersecurity risks associated with its suppliers and the products and services it acquires and uses

Table 3.16 Illustration of NIST CSF Tier 2

Tier	Cybersecurity risk government	Cybersecurity risk-management
Tier 2: Risk Informed	Risk-management practices approved by management but may not be established as organization-wide policy Prioritization of cybersecurity activities and protection needs directly informed by organizational risk objectives, the cyber threat environment, or business/mission requirements	Awareness of cybersecurity risks at organizational level, an organization-wide approach managing cybersecurity risks has not been established Consideration of cybersecurity in organizational objectives and programs may occur at some but not all levels of organization. Cyber risk assessment of organizational and external assets occur but is not typically, repeatable or reoccurring Cybersecurity information shared within organization on an informal basis Organization is aware of cybersecurity risks associated with its suppliers and products and services it acquires and uses, but it does not act consistently or formally in response to those risks

3.2 Cybersecurity Frameworks and Criteria

Table 3.17 Illustration of NIST CSF Tier 3

Tier	Cybersecurity risk government	Cybersecurity risk-management
Tier 3: Repeatable	Organization's risk management practices formally approved and expressed as policy Risk-informed policies, processes, and procedures defined, implemented as intended, and reviewed Organizational cybersecurity practices regularly updated based on application of risk-management processes to changes in business requirements, cybersecurity threats, and technological landscape	Organization-wide approach managing cybersecurity risks Cybersecurity information routinely shared throughout the organization Consistent methods are in place to respond effectively to changes in cyber risk Personnel possess know-ledge and skills to perform appointed roles and responsibilities Organization consistently and accurately monitors cybersecurity risks of assets Senior cybersecurity and non-cybersecurity executives communicate regularly regarding cybersecurity risks. Executives ensure that cybersecurity is considered through all lines of operation in the organization Organization risk strategy informed by cybersecurity risks associated with its suppliers, products and services, acquired and used. Personnel formally act upon those risks through mechanisms like written agreements to communicate baseline requirements, governance structures (e.g., risk councils), and policy implementation and monitoring. These actions are implemented consistently and as intended and continuously monitored and reviewed

Table 3.18 Illustration of NIST CSF Tier 4

Tier	Cybersecurity risk government	Cybersecurity risk-management
Tier 4: Adaptive	Organization-wide approach managing cybersecurity risks that uses risk-informed policies, processes, and procedures to address potential cybersecurity incidents Relationship between cybersecurity risks and organizational objectives is clearly understood and considered when making decisions. Executives monitor cybersecurity risks in the same context as financial and other organizational risks Organizational budget is based on an understanding of the current and predicted risk environ-ment and risk tolerance Business units implement executive vision and analyze system-level risks in the context of the organizational risk tolerances Cybersecurity risk management is part of the organizational culture. It evolves from awareness of previous activities and continuous awareness of activities on organizational systems and networks. Organization can swiftly and efficiently account for changes to business objectives in how risk is approached and communicated	Organization adapts its cybersecurity practices based on previous and current cybersecurity activities, including lessons learned and predictive indicators Through a process of continuous improvement that incorporates advanced cybersecurity technologies and practices, organization actively adapts to a changing technological landscape and responds in a timely and effective manner to evolving, sophisticated cyber threats Organization uses real-time or near real-time information to understand and consistently act upon cybersecurity risks associated with its suppliers and products and services it acquires and uses Cybersecurity information constantly shared throughout the organization and with authorized third parties

potential rewards expected from an action and subtracting the associated costs by turn data into smarter choices by the following steps:

- Establish the CBA framework for analysis.
- Identify all relevant costs and benefits.
- Quantify costs and benefits.
- Calculate net benefits and compare alternatives.
- Make recommendations based on findings.

3.2.2 MITRE ATTACK (MITRE ATT&CK)

MITRE ATT&CK® is a globally-accessible knowledge base of cyber attackers' tactics and techniques based on observations. The ATTACK knowledge base is used as a foundation for the development of specific cyber risk models and methodologies in the private sector, in government, and in the cybersecurity product and service community. With the creation of ATTACK, MITRE is fulfilling its mission to solve problems for a more cybersecure world by bringing communities together to develop more effective cybersecurity strategies. ATTACK is open and available to any organization or person for use at no charge [4]. The basis of ATTACK is the set of techniques and sub-techniques that represent actions that cyber attackers can perform to accomplish objectives. Those objectives are represented by the tactic categories, the techniques, sub-techniques and procedures. This relatively simple representation is a useful balance between sufficient technical detail at the technique level and the context around why actions occur at the tactic level. The main components of the model are [20]:

- **Tactics:** Represents *why* of an ATTACK technique or sub-technique. It is the cyber-attacker's tactical objective, the reason for performing an action. Tactics serve as useful contextual categories for individual techniques and cover standard notations for things cyber-attacker's do during an operation, such as persist, discover information, move laterally, execute files, and exfiltrate data. Tactics are treated as *tags* within ATTACK where a technique or sub-technique is associated or tagged with one or more tactic categories depending on the different results that can be achieved by using a technique.
- **Techniques:** Represents *how* cyber attacker's achieve tactical objective by performing an action. For example, a cyber attacker may dump credentials from an operating system to gain access to useful credentials within a network. Techniques may also represent *what* a cyber-attacker gains by performing an action. This is a useful distinction for the Discovery tactic as the techniques highlight what type of information a cyber-attacker is after with a particular action.
- **Sub-techniques:** Break down behaviors described by techniques into more specific descriptions of how behavior is used to achieve an objective. For example, with OS Credential Dumping, there are several more specific behaviors under this technique that can be described as sub-techniques.

- **Procedures**: Represent specific implementation cyber-attackers have for techniques or sub-techniques.

MITRE ATT&CK is organized in a series of technology domains, the ecosystem a cyber attacker operates within. Currently, there are three **technology** domains:

- **Enterprise:** Representing traditional enterprise networks and cloud technologies,
- **Mobile:** For mobile communication devices
- **ICS:** For industrial control systems

Within each domain are **platforms,** which may be an Operating System (OS) or application (e.g. Microsoft Windows). Techniques and sub-techniques can apply to multiple platforms.

MITRE ATTACK (MITRE ATT&CK) Use Cases

The MITRE framework includes a list of techniques that cyber attackers might use, such as spear phishing, command and control, and credential dumping. Thus, ATTACK use several cases for cyber-attack defense such as [20]:

- **Adversary Emulation:** Process of assessing cybersecurity of a technology domain by applying Cyber Threat Intelligence (CTI) about specific cyber-attackers and how they operate to emulate that cyber risk. Cyber-attacker emulation focuses on the ability of an organization to verify detection and/or mitigation of cyber attackers' activity at all applicable points in their lifecycle.
- **Behavioral Analytics Development:** Going beyond traditional Indicators of Compromise (IoCs) or signatures of malicious activity, Behavioral Detection Analytics (BAD) can be used to identify potentially malicious activity within network or information systems that may not rely on prior knowledge of cyber attackers' tools and indicators. It is a way of leveraging how a cyber attacker interacts with a specific platform to identify and link together suspicious activity that is agnostic or independent of specific tools that may be used.
- **Defensive Gap Assessment**: Allows an organization to determine what parts of its enterprise lack defenses and/or visibility. These gaps represent blind spots for potential vectors that allow cyber-attackers to gain access to its network and information systems undetected or unmitigated.
- **SOC Maturity Assessment:** An organization's Security Operations Center (SOC) is a critical component that continuously monitor for active cyber threat risk incidents against network and information systems. Understanding the maturity of a SOC is important to determine its effectiveness.
- **Cyber Threat Intelligence Enrichment:** CTI covers knowledge of cyber threat risk incidents and cyber threat actor groups that impact cybersecurity. It includes information about malware, tools, cyber attackers' Tactics, Techniques, and Procedures (TTPs), behavior, and other indicators that are associated to cyber threat risk incidents. TTP is an effective method for detecting malicious activity.

Table 3.19 MITRE ATT&CK development steps

Development step	Activity
1. Identify Behaviors	Identify and prioritize adversary behaviors from the threat model to detect
2. Acquire Data	Identify data that is necessary to detect a desired cyber attacker behavior. If capability to acquire data does not exist, a sensor must be created to collect this data
3, Develop Analytics	Create analytics from collected data to detect identified behaviors. It is also important to ensure that analytics do not have an unacceptable false positive rate on benign environmental events
4. Develop a Cyber Attacker Emulation Scenario	White Team develops a cyber attacker emulation scenario, based on ATTACK, that includes behaviors identified in Step 1
5. Emulate Threats	Red Team attempts to achieve the objectives outlined by the White Team by exercising behaviors and techniques described in the ATTACK model
6. Investigate Attack	Blue Team attempts to recreate the timeline of Red Team activity using analytics and data developed in Step 3 (Develop Analytics)
7. Emulate Performance	White, Red, and Blue Teams review the engagement to evaluate to what extent the Blue Team was able to use the analytics and sensor data to successfully detect the simulated APT behaviors. After this evaluation, the cycle repeats and returns to Step 1

More in general the ATTACK-based analytics development method contains seven steps that are shown in Table 3.19 [21]:

MITRE Engage: Cyber Denial, Deception, and Cyber-Attacker Engagement

Since cyber-attack techniques evolve continuously and be more sophisticated, cyber defense must also evolve to act at its best and sustain cyber defense. With it, new cyber defense solutions required, to counter todays sophisticated Tactics, Techniques and Practices (TTP) of cyber-attacks. In this regard, cyber defense must avoid, or at least fight through cyber degradation. Given that fact, the innovative approach of MITRE Engage is a framework for planning and discussing cyber-attacker's engagement measures that empowers the cyber defender to engage the cyber-attackers to implement cyber defenders high level cybersecurity goals [22]. MITRE Engage key concept is based on Cyber Denial, Deception, and Cyber-Attacker (Adversary) Engagement. As described in [23] Cyber Denial is the ability to prevent or otherwise impair cyber-attacker's ability to conduct their operations. This disruption may limit their movements, collection efforts, or otherwise diminish the effectiveness of their capabilities. In Cyber Deception cyber defender intentionally reveal deceptive facts and fictions to mislead the cyber-attackers. In addition, cyber defender conceals critical facts and fictions to prevent the cyber-attacker from forming correct estimations or taking appropriate actions. When cyber denial and deception are used together, and within the context of strategic planning and analysis, they provide the pillars of Adversary Engagement. In this regard, Deception technology is a category of cybersecurity solutions that detect cyber threat risk incidents early, with low rates of false positives. Doing so, deception technology deploys realistic decoys/honeypots in a network alongside real assets to act as lures. The moment a cyber-attacker interacts with a decoy, deception technology begins gathering important information. This information is used to generate high-fidelity alerts that reduce dwell time and speed up cyber threat risk incident response. Hence, the biggest benefit of deception technology is that it puts the burden of success on the cyber-attacker instead of the cyber defender. Once an entity is populated with decoys, cyber-attacker must carry out flawless cyber-attacks. This requires that cyber-attacker not falling for a single fake asset, misdirect, or trap, to succeed [24]. In this context, MITRE Engage mentioned in [23] that cyber-attacker's engagement is an iterative, goal driven process, not merely the deployment of a technology stack. It is not enough to deploy a decoy and declare success. Rather, organizations must think critically about what their defensive goals are and how Denial, Deception, and Cyber-Attacker Engagement can be used to drive progress towards these goals. Unlike other defensive technologies, such as antivirus, Cyber-attacker Engagement technologies cannot be considered as fire and forget solutions. Thus, cyber defenders still needed to think carefully about their goals when designing their Deception technology engagement. Therefore, MITRE Engage was created to support the private sector, government, and vendor communities plan and execute cyber-attacker's engagement activities by providing a shared reference to bridge the gaps across these communities. In this regard, a core offering of MITRE Engage is the Engage Matrix, which was built based on MITRE expertise in adversary engagement and our knowledge of adversary behavior observed in the real world. The Engage Matrix is intended to enable the discussion and planning of cyber-attacker Engagement, Deception, and Denial activities to drive strategic cyber outcomes [23]. For this purpose, the Engage Matrix is divided into two categories of actions. Strategic actions bookend the matrix and ensure that cyber defenders appropriately drive operations with strategic planning and analysis. Additionally, the strategic actions ensure that the cyber defender has gathered appropriate stakeholders (management, legal, etc.) and defined acceptable risk. Throughout the operation, the cyber defender will ensure that all engagements operate within these defined guardrails. Engagement actions are the traditional Cyber Denial and Deception activities that are used to drive progress towards cyber defender's objectives [23]. With Cyber Denial, Deception and Counter Deception, based on MITRE research, an active novel cyber defense framework is available [25].

3.2.3 CIS Critical Security Controls

The Center for Internet Security (CIS) assembled a working group of policy experts to develop the *Information Security Policy* templates. These policy templates align with CIS Controls version 8. At a high level, the CIS Controls are best practice recommendations that consist of a prioritized set of actions to defend against the most common cyber threat risk incidents. In version 8.1 of the Controls, there are 18 *controls*, followed by a subset of 153 *actions* called Safeguards [26, 27]. To support with prioritization, the Safeguards are divided into three Implementation Groups (IGs): IG1, IG2, and IG3 [27].

- **IG1:** Essential to cyber hygiene and represents the minimum standard of information security for all organizations These are the actions that every organization should take first, regardless of size. IG1 contains 56 cyber threat risk incident defense safeguards.
- **IG2:** Assists organizations managing the IT infrastructure of multiple departments with differing cybersecurity risk profiles. It aims enabling organizations to cope with increased operational complexity. IG2 contains 74 additional cyber threat risk incident defense safeguards.
- **IG3:** Assists organizations with IT security experts to cybersecure sensitive and confidential data. It aims to prevent and/or lessen the impact of sophisticated cyber threat risk incident attacks. IG3 contains 23 additional cyber threat risk incident defense safeguards.

IGs are based on several factors: *Size and/or Complexity*, *Data Types*, *Resources and Technology*, *Threat Types*, and *Risk*. Each IG identifies a set of Safeguards that the enterprise should implement.

The size, e.g. employees, and complexity, e.g. operational-wise, factors that may impact the IG selection is illustrated in Table 3.20 [28].

The types of data an organization deal with and stores has also impact which IG is selected, as illustrated in Table 3.21 [28].

The resources, e.g. budget, skill sets, and others, and technology, e.g. existing or no-cost technology, Commercial Off-The-Shelf (COTS) products, or use proprietary/custom-built software, may impact the IG selection is illustrated in Table 3.22 [28].

There are different types of cyber threat risk incidents that an organization can be exposed to like ransomware, malicious code and others. For the CIS Controls, the Safeguards are primarily focused on cyber threat risk incidents. Cyber threat risk incident can be opportunistic cyber-attacks, where the cyber-attacker takes advantage of an opportunity but is not necessarily targeting an organization. They may also be targeted and planned well in advance of the actual cyber threat risk incident, aiming elicit a specific result. Cyber threat risk incidents impacting the IG selection are illustrated in Table 3.23 [28].

Risk-management is at the center of most organizations' business operations, but there are different facets of risk-management. Two key components that often factor into the selection of IGs are risk tolerance and risk willingness. In the realm of cyber threat risk incidents, an organization may have a risk willingness for service disruptions to their website, but one of the risk tolerance criteria indicates that the organization's website can be down no longer than some hours. Thus, risk factors impacting the IG selection are illustrated in Table 3.24 [28].

3.2 Cybersecurity Frameworks and Criteria

Table 3.20 Size and/or complexity factors impacting IG

IG	Size and/or complexity of organization
IG1	Small and medium-sized organization Starting point for all enterprises
IG2	Enterprises managing IT infrastructure spanning multiple departments with differing cybersecurity risk profiles Increased operational complexity
IG3	Large Organizations that are operationally complex Increased cybersecurity risk profile across the organization

Table 3.21 Data type factors impacting IG selection

IG	Data type factors
IG1	Data is low sensitivity Stores unregulated employee and organizations financial information No regulatory or compliance oversight
IG2	Stores and processes sensitive client or organization information Pockets of regulatory or compliance oversight
IG3	Stores and processes sensitive and confidential data Subject to regulatory and compliance oversight

Table 3.22 Resources and technology factors impacting IG selection

IG	Resources and technologies
IG1	Limited cybersecurity expertise, may outsource IT/cyber-security needs Use existing and/or cost-effective technology and processes Use Commercial off-the-Shelf (COTS) products
IG2	Implements organization-grade technology Has specialized expertise to install and properly configure software Use COTS products, and may using proprietary/in-house software
IG3	Employs cybersecurity experts that specialize in various areas of cybersecurity Likely using software designed in-house or proprietary to the organization, in addition to COTS products Ability to purchase dedicated cybersecurity software for specific tasks

Table 3.23 Cyber threat risk incidents impacting IG selection

IG	Cyber threat risk incidents
IG1	Exposed to general, non-targeted attacks, e.g., Ransomware, Malware, Web Application Hacking, Insider and Privilege Misuse
IG2	Exposed to more advanced cyber threat risk incidents, resulting in a loss of public trust Exposed to industry-specific cyber threat risk incidents
IG3	Exposed to sophisticated and targeted cyber threat risk incident attacks, including zero-day attacks and/or nation-state cyber threat risk incident actors Successful cyber threat risk incident attacks can cause significant harm to the public welfare

Table 3.24 Risk factors that impact IG selection

IG	Risk factors
IG1	Limited tolerance for downtime Focus is to keep business operational
IG2	Faces an increased risk exposure (probability x potential losses) Can withstand short interruptions of service Concern is loss of public trust if a breach occurs
IG3	Faces the highest risk exposure Cannot withstand interruptions of service

Each Safeguard is also assigned an *Asset Class*, defined as a group of information assets that are evaluated as one set based on their similarity. In v8.1, Asset Classes are categorized into *Devices, Software, Data, Users, Network,* and Documentation. Within each Asset Class, there are a number of sub-categories that align to language used throughout the CIS Controls [27].

Furthermore, CIS Controls has developed information security policy templates. These policy templates align with CIS Controls v8 and v8.1, enabling organizations to formalize their efforts around addressing the Safeguards in Implementation Group 1 (IG1). They cater exclusively to IG1 Safeguards; they do not address Implementation Group 2 (IG2) or Implementation Group 3 (IG3) Safeguards. The templates are shown in Table 3.25 [27].

CIS Controls actual version 8.1 is an iterative update to version 8.0 that evolve the CIS Controls, establishing *design principles* that guide through any minor or major updates. The design principles for v8.1 are context, clarity, and consistency [29].

- **Context:** Enhances CIS Controls with new asset classes to better match the specific parts of an organization's infrastructure to which each CIS Safeguard applies. New classes require new definitions, so the descriptions of several CIS Safeguards also enhance for greater detail, practicality, and clarity.
- **Clarity:** Aligns with other major cybersecurity frameworks to the extent practical, while preserving the unique features of the CIS Controls.
- **Coexistence:** CIS Controls always aligned to evolving industry standards and frameworks, and will continue to do so. This is a core principle of how the Controls operate; it assists all users of the Controls. The release of NIST CSF 2.0 necessitated updated mappings and updated cybersecurity functions.
- **Consistency:** Any iterative update to the CIS Controls should minimize disruption to Controls users. Hence, no IGs were modified in this update, and the context of any given CIS Safeguard remains the same. Additionally, new asset classes and definitions needed to be consistently applied throughout the Controls, and in doing so, some minor updates were added, as covered in the CIS Controls v8.1 to the following updates:

  – Realigned NIST CSF cybersecurity function mappings to match NIST CSF 2.0,

3.2 Cybersecurity Frameworks and Criteria 187

Table 3.25 Policy templates

Template	Policy template factor
1	Acceptable Use Policy Template for the CIS Controls Template can assist an organization in developing acceptable use for the CIS controls
2	Enterprise Asset Management Policy Template for CIS Control 1 Template can assist an organization in developing an asset management policy
3	Software Asset Management Policy Template for CIS Control 2 Template can assist an organization in developing a software asset management policy
4	Data Management Policy Template for CIS Control 3 This template can assist an enterprise in developing a data management policy
5	Cybersecure Configuration Management Policy Template for CIS Control 4, 9, and 12 Template can assist an organization in developing a cybersecure configuration management policy
6	Account and Credential Management Policy Template for CIS Controls 5 and 6 Template can assist an organization in developing an account and credential management policy
7	Vulnerability Management Policy Template for CIS Control 7 Template can assist an organization in developing a data management policy
8	Audit Log Management Policy Template for CIS Control 8 Template can assist an organization in developing an audit log management policy.
9	Malware Defense Policy Template for CIS Control 10 Template can assist an organization in developing a malware defense policy.
10	Data Recovery Policy Template for CIS Control 11 This template can assist an enterprise in developing a data recovery policy.
11	Cybersecurity Awareness Skills Training Policy Template for CIS Control 14 Template can assist an organization in developing a cybersecurity awareness skills training policy
12	Service Provider Management Policy Template for CIS Control 15 Template can assist an organization in developing a service provider management policy
13	Incident Response Policy Template for CIS Control 17 Template can assist an organization in developing an incident response policy

- Included new and expanded glossary definitions for reserved words used throughout the Controls, e.g., plan, process, sensitive data,
- Revised asset classes alongside new mappings to CIS Safeguards,
- Fixed minor typos in CIS Safeguard descriptions,
- Added clarification to a few anemic CIS Safeguard descriptions.

CIS Controls v8.1 also added the *Governance* cybersecurity function. So, organizations can't steer their cybersecurity program toward achieving their goals without the structure provided by effective governance. CIS Controls v8.1 specifically identifies governance topics as recommendations that organizations can implement to enhance their governance of their cybersecurity program.

Table 3.26 CIS Critical Security Controls

CIS controls	Alignment
Control 1	Inventory and Controls of Organizations Assets
Control 2	Inventory and Control of Software Assets
Control 3	Data Protection
Control 4	Secure Configuration of Organizations Assts and Software
Control 5	Account Management
Control 6	Access Control Management
Control 7	Continuous Vulnerability Management
Control 8	Audit Log Management
Control 9	Email and Web Browser Protection
Control 10	Malware Defenses
Control 11	Data Recovery
Control 12	Network Infrastructure Management
Control 13	Network Monitoring and Defense
Control 14	Cybersecurity Awareness and Skills Training
Control 15	Service Provider Management
Control 16	Application Software Cybersecurity
Control 17	Incident Response Management
Control 18	Penetration Testing

CIS Controls v8.1 also aim to streamline the process of designing, implementing, measuring, and managing organizations cybersecurity. This involves simplifying language to reduce duplication, focusing on measurable actions with defined metrics, and ensuring each CIS Safeguard is clear and concise. CIS continues to balance the need to address current cybersecurity challenges and maintain a stable, foundational cybersecurity defense strategy, all the while steering clear of overly complex or inaccessible technologies. Technology is constantly shifting, and CIS Controls team is aware of the developments in Artificial Intelligence (AI), Augmented Reality (AR), and ambient computing working to reshape organizations infrastructure in subtle and radical ways. Thus, CIS Controls team is aware about and already working on ideas for Version 9 of the CIS Controls.

CIS Critical Security Controls v8.1 contain 18 Controls, shown in Table 3.26 [30].

3.2.4 ISO/IEC 27K

ISO/IEC 27K is a family of international standards developed by the International Organization for Standardization (ISO) and the International Electrotechnical Commission (IEC) that specify requirements and guidelines for establishing, implementing, maintaining, and improving an Information Security Management, vital for organizations. ISO/IEC 27001 is a well-known standard for information security management and their requirements, as described in the ISO/IEC 27001:2022 Information security, cybersecurity and privacy protection framework. Hence, the ISO/IEC 27001:2022 standard provides organizations of any size and from all sectors of activity with guidance for establishing, implementing, maintaining and continually improving an Information Security Management System (ISMS) to manage cyber threat risk incidents related to the organizations cyber risk.

With the ever evolving and sophisticated cyber-crime **new cyber threat risk incidents constantly emerging,** and it's difficult or even impossible to manage the manifold cyber risks. By that, ISO/IEC 27001:2022 enables organizations become cyber risk-aware and proactively identify and address organizations weaknesses. ISO/IEC 270012022 promotes a holistic approach to information cybersecurity by vetting people, policies and technology. An ISMS implemented according to this standard is a tool for **risk-management, cyber-resilience** and **operational excellence. Against this background,** an ISMS incorporates features on which experts in the field of cybersecurity have reached a consensus being the international state of the art. Therefore, ISO/IEC JTC 1/SC 27 maintains an expert committee dedicated to the development of international management systems standards for information security, also known as ISMS family of standards [6].

Through the use of the ISMS family of standards, organizations can develop and implement a framework for managing cybersecurity of their information assets, including financial information, intellectual property, and employee details, or information entrusted to them by customers or third parties. These standards can also be used to prepare for an independent assessment of their ISMS applied to them protection of information. An ISMS is a systematic approach to managing the CIA triad (see Sect. 1.5) of data and assets. In this context ISO/IEC 27K includes more than 40 standards. For example, ISO/IEC 27001 and ISO/IEC 27002 are relevant to cybersecurity, whereby the ISO/IEC 27001 defines the requirements for an ISMS, while ISO/IEC 27002 provides a code of practice for information cybersecurity controls. Thus, the content of the ISO/IEC 27K can be summarized by the five subsequent sections, shown in Table 3.27 [6].

For Section 5 in Table 3.27 the ISO/IEC 27K family is divided into sub-categories, as shown in Table 3.28. Together, they enable organizations of all sectors and sizes to manage cybersecurity of assets such as financial information, intellectual property, employee data and information entrusted by third parties.

The ISO/IEC 27001:2022 requirements in Table 3.29 illustrates the essential Annexes of the standard.

Table 3.27 Contents of ISO/IEC 27K

ISO/IEC 27K Sections	Content
1	Scope
2	Normative References
3	Terms and Definitions
4	Information Security Management Systems (ISMS)
5	ISMS Family Standards

Table 3.28 ISO/IEC Standards used in Section 5 in Table 3.29

Section 5 ISO/IEC 27K	ISO/IEC 27K Standard
Standards specifying requirements	ISO/IEC 27001
	ISO/IEC 27006
	ISO/IEC 27009
Standards describing general guidelines	ISO/IEC 27002
	ISO/IEC 27003
	ISO/IEC 27004
	ISO/IEC 27005
	ISO/IEC 27007
	ISO/IEC TR 27008
	ISO/IEC 27013
	ISO/IEC 27014
	ISO/IEC TR 27016
	ISO/IEC 27013
Standards describing sector specific guidelines	ISO/IEC 27010
	ISO/IEC 27011
	ISO/IEC 27017
	ISO/IEC 27018
	ISO/IEC 27019
	ISO 277799

Table 3.29 ISO/IEC 27001:2022 requirements annexes

ISO/IEC 27001 Requirements	ISI/IEC 27001:2022 Annex Title
Annex 4.1	Understanding the organization and its context
Annex 4.2	Understanding the needs and expectations of interested parties
Annex 4.3	Determining the scope of the ISMS
Annex 4.4	ISMS
Annex 5.1	Leadership and commitment
Annex 5.2	Information security policy
Annex 5.3	Organizational roles, responsibilities and authorities
Annex 6.1	Actions to address risks and opportunities
Annex 6.2	Information security objectives and planning to achieve them
Annex 7.1	Resources
Annex 7.2	Competence
Annex 7.3	Awareness
Annex 7.4	Communication
Annex 7.5	Documented information
Annex 8.1	Operational planning and control
Annex 8.2	Information security risk assessment
Annex 8.3	Information security risk treatment
Annex 9.1	Monitoring, measurement, analysis and evaluation
Annex 9.2	Internal audit
Annex 9.3	Management review
Annex 10.1	Nonconformity and corrective action
Annex 10.2	Continual improvement

3.2 Cybersecurity Frameworks and Criteria 191

Table 3.30 ISO/IEC 2700:2022 controls annexes

ISO/IEC 27001 Controls	ISI/IEC 27001:2022 Annex Title
Annex 5	Information security policies
Annex 6	Organization of information security
Annex 7	Human resource security
Annex 8	Asset management
Annex 9	Access control
Annex 10	Cryptography
Annex 11	Physical and environmental security
Annex 12	Operations security
Annex 13	Communications security
Annex 14	System acquisition, development, and maintenance
Annex 15	Supplier relationships
Annex 16	Information security incident management
Annex 17	Information security aspects of business continuity management
Annex 18	Compliance

The ISO/IEC 27001:2022 controls in Table 3.30 show the essential Annexes of the standard.

3.2.5 *Difference Between NIS CSF 2.0 and ISO/IEC 27K*

NIST CSF 2.0 and ISO/IEC 27K have similarities that bring several benefits to cybersecurity planning. Both, framework and standard, are based on risk-management principles that are international standards and best practices, which can be tailored to any organization, regardless of size, industry or region. Continuous improvement cycle (not one-time compliance approach) to regularly review and update cybersecurity policies, procedures and practices.

Despite similarities, there are significant differences between NIST CSF 2.0 and ISO/IEC 27K series that may influence the selection and implementation of each framework or standard. For example, NIST CSF 2.0 is a framework, not a standard that focuses on cybersecurity, not information security. Structured around six core functions and is voluntary. In contrast, ISO/IEC 27K series is a set of standards that specify requirements and guidelines to build up an Information Security Management System (ISMS) and related controls. It covers broader aspects of information security, embedded around ISMS and is certifiable. NIST CSF 2.0 does not require formal assessment or verification of conformity, it relies on organizations' self-assessment and self-improvement. ISO/IEC 27K series enables independent certification of an ISMS by an accredited external auditor, providing a higher level of assurance and recognition.

3.2.6 Maturity Models After ISO 9004:2008/ 2015

Digital transformation significantly improves organizations process performance and their innovation. For sustainable organizational success, the ISO 9004:2008/2015 contains a self-assessment tool to identify organizations strengths and weaknesses to cybersecurity issues, and determine and define prioritized measures for improvement. The measured results and/or potentially progress of the cybersecurity maturity is illustrated graphically, using a radar chart representation, shown in Fig. 3.4, also called spider diagram. The diagram represents the strength and weakness of essential features of organizations, expressed by the respective maturity level. The diagram axes are linked, the graph is divided into multiple grids that represent the interesting information. Figure 3.4 show five diagrams with maturity levels from 3 to 7, represented by shown pentagons.

Success is provided by self-assessment, carried out regularly and by improvements. The elements of ISO 9004:2008/2015 assigned graded requirements that enable identifying the current assessment status and possible improvements. Compared to generic maturity models, application-specific maturity models have the advantage of precise measurement of the improvement status due to the specific indicators. ISO 9004:2008/2025 includes specific features and capabilities [32, 33], as shown in Table 3.31:

The structure in Table 3.31 enable improving a sustainable maturity development model based on quality management standards, because many considerations on maturity successfulness include elements of a Quality Management System (QMS). A QMS is a tool that effectively controls, develops, and monitors activities with different focus areas, e.g., producing products focusing on high quality, taking care of the environment, keeping information cybersecure, or creating a good working environment. Therefore, ISO 9004:2008/2015 is guidance for achieving sustained success for assessment of a maturity level. For example, the PhD thesis in [34] describes a maturity model assigned to five maturity levels, using defined criteria of ISO 9004:2008/2015.

For the ISO 9004:2008/2015 Management towards a Sustainable Organizations Success, a digital management approach update, a model for self-assessment of organizations maturity with respect to key elements of organizations functioning, is published in [35]. This publication assigned the key elements in the following areas, focusing on suitable organizations success strategy and policy, resource and process management, monitoring, measurements, analysis and review improvement innovation and learning.

Fig. 3.4 Spider diagrams with different maturity levels from 3 to 7 [31] with 5 equilateral triangles, called dimensions

3.2 Cybersecurity Frameworks and Criteria 193

Table 3.31 ISO 9004:2008/2015 Features and capabilities

Criteria no.	Features	Capability
I	Foreword	./.
II	Introduction	./.
1	Scope	Give guidelines enhancing organizations ability to achieve sustained success. Guidance is consistent with quality management principles given in ISO 9004:2008/2015. Document provides a self-assessment tool to review the extent to which the organization has adopted the concepts. Document is applicable to any organization, regardless of size, type and activity
2	Normative references	The documents referred to is in such a way that some or all of their content constitutes requirements of this document. For dated references, only the edition cited applies. For undated references, the latest edition of the referenced document (including any amendments) applies
3	Definition of sustainable development	3.1 Introduction 3.2 Organization is able to maintain and develop its performance over a longer period of time
4	Management of sustainable development	4.1 Introduction 4.2 Key features of sustainable development (organization) 4.3 Management of sustainable development 4.4 Evaluation of sustainable development
5	Environment Organization	5.1 Introduction 5.2 Monitoring 5.3 Analysis

6	Strategies, policies and communication	6.1 Strategic orientation
		6.2 Mission and vision
		6.3 Aspects of the strategy
		6.4 Policies and objectives
		6.5 Strategic planning
		6.6 Risk management
		6.7 Review strategies for sustainable development
		6.8 Communications
		  6.8.1 General
		  6.8.2 Effectiveness and efficiency of the process of communication

7	Resources	7.1 Resource management
		7.2 Planning
		7.3 Resource allocation
		7.4 Resources staff
		  7.4.1 General
		  7.4.2 Motivation and involvement of employees
		7.5 Infrastructure
		  7.5.1 General
		  7.5.2 Working conditions
		7.6 Knowledge
		7.7 Financial resources
		7.8 Natural resources management and life expectancy

(continued)

Table 3.31 (continued)

Criteria no.	Features	Capability
8	Processes	8.1 Process approach 8.2 Types of processes 8.3 Management processes of the organization 8.4 Responsibility and authority for the process
9	Measurement and Analysis	9.1 Measurement approach 9.2 Performance matrix 9.3 Measurement of achieving the goals 9.4 Key indicators 9.5 Tools for measuring 9.6 Internal audit 9.7 Evaluation 9.8 Review and evaluation process

10	Learning	10.1 Introduction
		10.2 Learning
		  10.2.1 Types of learning
		  10.2.2 Sources of learning
		  10.2.3 Factors that influence the effectiveness of learning
		  10.2.4 Planning of learning
		10.3 Improvement
		10.4 Innovation
		  10.4.1 Generally
		  10.4.2 Types of innovation
		  10.4.3 Factors that influence the effectiveness of innovation
		  10.4.4 Planning process innovation

Appendix A	Tools for assessing the maturity	A.1 Introduction A.2 Description of the level of maturity A.3 Assessment Strategy A.4 Mark work A.5 Tools for evaluation A.6 Results evaluation and improvement planning
Appendix B	Forms estimation of the maturity	./.

Another ISO 9004:2008/2015 maturity model published focus on building optimal management for the maturity level of organizations business processes, whereby the level of maturity describes a linear dynamic control system [36]. Developing the model, the method of Analytical Construction of Optimal Regulation (ACOR) is used. The constructed model shows plausible behavior in predicting the process of managing organizations maturity and reproduces the effect of accelerating growth of controlled indication, identified in the model as a-priority. As published in [37] organizations may not aware or not capable of assessing their Industry 4.0 maturity level. They used a maturity model based on ISO 9004:2018 for assessment of maturity levels of quality in Industry 4.0, framed as Quality 4.0, developed based on a questionnaire with a sample size of 335 companies to explore the topic. Confirmatory Factor Analysis (CFA) and Security Event Management (SEM) used to verify the ISO 9004:2018 model potential, used for assessing Quality 4.0 maturity level. CFA is a statistical technique used to verify the factor structure of a set of observed variables. It allows to test the hypothesis that a relationship between observed variables and their underlying latent constructs exists. CFA is distinct from Exploratory Factor Analysis (EFA), where the structure of the data is not predefined and is instead determined through the analysis. SEM is a statistical approach that enables to explore and analyse the relationships between observed variables and underlying latent constructs. It effectively combines principles from factor analysis that identifies underlying factors from observed variables, and multiple regression analysis, which assesses how one set of variables predicts another. In addition, resource management proved to have a positive impact on performance analysis and evaluation as well on improvement, learning and innovation constructs. The findings indicate the model is usable in the context of Quality 4.0, and is utilized as a basis for developing a sustainable Quality 4.0 system roadmap.

3.2.7 Network and Information System Security Directive (NIS2)

The European Union (EU) published end of 2022 the document "Measures for a high common level of cybersecurity across the Union," called NIS2, which is the successor of the "Network and Information Security Directive" (NIS), released in 2016. The NIS2 directive provides legal measures to improve cybersecurity across the EU Member states, including obligations for businesses, government authorities, and cooperation between EU Member States on cyber risks. NIS2 will become effective across all EU Member states by end of March 2025. All requirements apply for entities of high critical and critical sectors and business sizes that are active in one or more EU member states. Entities in scope with NIS2 has to engage in economic activities, irrespective of its legal form. Major suppliers for clients in the EU Member states will also be in scope if their clients are in the EU Member states and receive critical services. All regulative and practical NIS2 measures are described in detail in Chap. 2. Based on Chap. 2 key features of NIS2 are:

- **Achieve Level of Reliability:** Fines and penalties for repeated non-compliance are intended, up to temporary suspension of C-Level/board members.
- **Enhanced Sectors:** Supervision and enforcement by local authorities for medium and large organizations across high critical and critical sectors.
- **Minimum Requirements:** Establish mandatory, reviewable, and sanctionable cybersecurity measures for potential cyber threat risk incident reporting and risk-management but not only covering cyber threat risk incident detection but also taking response/remediation actions, and measures the outcome and cybersecurity management efficiency.
- **Supply Chain:** Requires risk reviews of cybersecurity practices for major connected 3rd party services providers, e.g., managed cybersecurity services providers or cybersecurity of pre-processing companies.

Expected Impacts by NIS2 are:

- **Cybersecurity Maturity insights across EU Member states:** Develop a good understanding of effective measures and raise the cybersecurity maturity to an expected common level.
- **Efficient Risk-Management:** Adding active/proactive risk-management measures that add cost/time efficiency pressure to organization as it requires not only to detect but to continuously assess and respond to risks.
- **NIS2 and more:** Cybersecurity maturity is a strategic target of the EU commission. Parallel to NIS2, additional initiatives are underway.

Suggested Activities to NIS2 are:

- **Review Compliance**: Assess cybersecurity risks based on all relevant organization's assets, review cybersecurity risk-management and cyber threat risk incident detection as well as response management capabilities, and establish local/regional/global EU responsibilities.
- **Close Gaps:** Identify OT-accepted cybersecurity solutions based on their cybersecurity coverage, efficiency, and operational costs. Use a cybersecurity demo to run a pilot before selecting a vendor. There are OT security solutions for different maturity levels available. Choose one that covers regulatory compliance requirements as well as internal ones. Keep in mind the need to increase for a higher maturity level over time, make sure the investment is sustainable, and the solution doesn't have to be replaced swiftly because requirements may have increased.
- **Note to CISOs and Board Members:** Be aware of what's coming, because CISOs and Board Members can be held personally responsible, and also, because the EU is building and enforcing a major and rather complex framework of cybersecurity compliance regulations across all EU Member states.

Other Europe EU directives and laws beside NIS2 are the Critical Entity Resilience (CER) that lays down obligations on EU Member States to take specific measures, to ensure that essential services for the maintenance of vital societal functions or economic activities are provided in an unobstructed manner in the internal market; and the EU Cyber Resilience Act (ECRA) that aims to safeguard consumers and businesses buying or using products or software with digital components. CER and CRA have significant consequences for all involved, vendors, operators, and service providers of operational equipment.

All regulative, practical and technological measures, essential to NIS2, described in detail in Chap. 2.

3.3 Cybersecurity Maturity Models

Data analytics of cyber risks analytics are a central feature of obtaining in depth knowledge of potential vulnerabilities and cyber risk of organizations crucial and critical digital data, assets, processes, systems, services and others. Therefore,

3.3 Cybersecurity Maturity Models

evaluating the *Current* Maturity Profile of cybersecurity capture the situational cyber risk situational awareness, an increasing part in active and/or proactive cybersecurity plans, to earn an adequate cybersecurity maturity level, defined as *Target* Maturity Profile. The *Target* Maturity Profile should always be higher then the *Current* Maturity Profile. With it, the cybersecurity maturity level model is a continuous process improvement, based on many small, evolutionary steps rather than revolutionary innovations, organized by these evolutionary steps with their impact. The steps define an ordinarily scale to measure the cybersecurity maturity level of an organization, a well-defined evolutionary plateau toward achieving a mature cybersecurity process that meets the targeted maturity objectives. As described in [38], the maturity level measuring method calculates the weightages of the maturity of all cybersecurity control objectives defined under the different control domains, whereby the criticality of each control objective depends of the organizations business functionality. Hence, the Cybersecurity Maturity Level (CML) of a business function can be calculated as follows [38]:

$$CMLBi = \frac{\sum_{j=1}^{n}(MLC_j * WC_j)}{\sum_{j=1}^{n}(WC_j)}$$

with

$CMLB_i$: Cybersecurity maturity level of business functions i where $i = 1$ to n,
n: Total number of business functions,
MLC_j: Cybersecurity maturity level of control j,
WC_j: Cybersecurity control weightage of control j,
K: Total number of controls identified.

In this regard, a cybersecurity maturity model is a framework of cybersecurity practices, guidelines, and controls, provide an organization with a proactive planning for creating effective, and at times compliant, cybersecurity programs.

3.3.1 Cybersecurity Maturity Level Model

The Cybersecurity Maturity Level Model (CMLM), evaluates organizations cybersecurity processes, systems, and service performance, to estimate their potential cybersecurity risk(s). However, it's more than just technological aspects [37, 39]. No uniform model exists to determine the cybersecurity maturity level of an organization. The technological dimension is related to the current situation to cybersecurity awareness of cybersecurity processes, systems, and others. Supervene issues are associated with critical issues to organizations cybersecurity, such as: i) employees with the necessary know-how and skills, and ii) necessary and approved budgets [40, 41].

Applying CMLM, the underlying dimensions must be operational by scaling the evaluation criteria for organizations required cybersecurity maturity levels. An example is shown in Table 3.32 that scaled the evaluation criteria, for example, in six maturity levels from 0 to 5.

The scale in Table 3.32 and the assignments of the evaluation criteria be usually illustrated in a radar chart as shown in Fig. 3.5. This 2-D chart displays multi-variate data of quantitative variables, stacked at an axis with the same central point, and plots the data as a polygonal shape over all axes. The chart features two or more quantitative variables for comparison, which are known as radii. To create a radar chat, the chart elements chart instance and data are needed, at least two axes: radial that spans from the center of the radar circle and circular that goes around the outer perimeter of the radar circle. Radar charts plots a series of observations or cases with multivariate data for comparison purposes, whereby a polygon represents each observation or case. However, graphing multiple observations or cases can become messy, because it can be hard to distinguish more than three or more stacked polygons visually. In general, the map looks similar to the spider web, which is why it's also called a spider chart. Often the data values of the radar chart are mapped using transparent colors, shades, patterns, dashed lines, bold lines and others, to illustrate differences and similarities instantly easier, and make it simple showing the variation between various data points.

Interpreting the representation of cybersecurity maturity levels of the investigated objectives be shown for five of the six NIST CSF 2.0 core functions in the radar chart in Fig. 3.2. It illustrates that non-of-them achieve a maturity level of 3 and/or higher. Maturity levels below 3 represent the perseverance of one or more significant obstacles to attaining the objective of Level 3. This means for the results shown in Fig. 3.5 that, for example, the cybersecurity personal under consideration in the respective organizations don't have sufficient cybersecurity knowledge and skills to support five NIST CSF 2.0 core functions adequately. Hence, the CMLM

Table 3.32 Maturity levels with assessment criteria

Maturity level	Assessment criteria
0	No activities toward situational cybersecurity awareness and cybersecurity risk assessment
1	Planning on the subjects of situational cybersecurity awareness and cybersecurity risk assessment, but no concrete implementation yet
2	Parts of planning situational cybersecurity awareness and cybersecurity risk assessment have already been implemented
3	Topic situational cybersecurity awareness and cybersecurity risk assessment is fully implemented and fully documented
4	Topic situational cybersecurity awareness and cybersecurity risk assessment is continuously checked for the state of the art and efficiency
5	Topic situational cybersecurity awareness and cybersecurity risk assessment is subject to a continuous improvement process

3.3 Cybersecurity Maturity Models

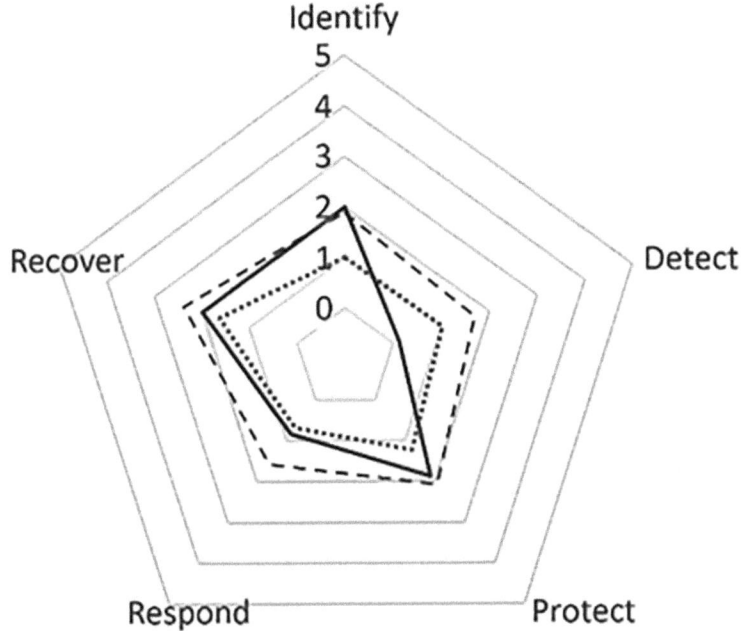

Fig. 3.5 Radar chart pentagon with 5 equilateral triangles, called dimensions, that, e.g., represent 5 of the six core functions identify, detect, protect, respond, and recover of the NIST Cybersecurity Framework (NIST CSF 2.0). The 5 equilateral triangle sites are divided into 6 maturity levels, ranging from 0 to 5, that compares the maturity of different investigated objectives, e.g., employee skills (dotted line), cybersecurity plan (bold line), cybersecurity risk management (dashed line)

shows the objectives to be further developed that need a respective action to lift them at least to the mean maturity level 3 [42].

Another condition, determining the level of cybersecurity maturity, is to estimate the effort that arises from the NIST CSF 2.0 criteria for a higher maturity level by eliminating the identified weaknesses. In this regard, the *Current Profile* maturity level is merely a part of an intermediate state to determine the knowledge, skills or activities to acquire the next higher maturity level. Against this background, CMLM also initiate recommendations for further development of the *Current Profile* in the direction of the *Target Profile* as to be state of the cybersecurity capability.

Beside the foregoing illustration of the CMLM, applications can also be more specifically to generate a cybersecurity maturity model, for example, for a maturity level of cybersecurity awareness culture, as shown in Table 3.33.

From Table 3.33 it can be seen that cybersecurity maturity level 5 depict a very low cyber risk. Low cybersecurity risk means there are only few anomaly cases referred to as cyber risks, outside a general concern for probable cyber threat risk incidents. Therefore, at low risk for cyber-breaches likelihood the organizations cybersecurity process, system and service activities deemed well positioned, which mean no major or dangerous incidents are to be expected.

Table 3.33 CMLM criteria depicted for maturity of cybersecurity awareness culture

Maturity level	Maturity level assignment	Maturity level criteria
1	Basic Compliance	No real cybersecurity awareness culture activities; high risk for severe cyber-break likelihood
2	Basic Security Awareness	Basic cybersecurity awareness culture; risk for cyber-break likelihood are moderately difficult
3	Programmatic Security Awareness & Behavior	Cybersecurity awareness culture programmatic wise already implemented; risk for cyber-break likelihood is medium
4	Security Behavior Management	Cybersecurity awareness culture is implemented & fully documented; risk for cyber-break likelihood persist at light
5	Sustainable Security Culture	High cybersecurity awareness culture; low risk for cyber-break likelihood

Despite of low cyber threat risk incident likelihood, cybersecurity tools and methods required to defend a potential successful execution outside the usual concern for cyber threat risk incidents. This can be achieved by using cyber risk-management, cyber compliance-management, business continuity-management, policy and documentation-management, to defend potential vulnerabilities, misuse of passwords, insider attacks, malicious code intrusion and others to maintain a sustainable and high cybersecurity maturity level in cybersecurity processes, systems, service and others.

The aim of CMLM is to illustrate which activities of the organization plans are still inadequately and where are need sfor action. If the level of cybersecurity maturity is low, a greater need for adaptation is required than with a higher level of maturity. Thus, the cybersecurity maturity model can be classified as an evaluation model for representing both current entrepreneurial technological and economic measures along the application reference, based on a reflexive self-assessment. For this purpose, a priori knowledge of the company's current situation is essential, which may need to be adapted to specific framework conditions. The outcome obtained by this process result in an improved cybersecurity maturity model, as illustrated in the radar chart.

Related to NIST CSF 2.0 core function *Protect* in Fig. 3.5, questions can be worked out, regarding the *Current Profile* of technical measures such as network segmentation, separation of network segments through firewall functions, and others. The NIST CSF 2.0 core function *Protect* includes not only pure technical measures, but also procedural ones like regular implementation of cybersecurity training for employees beside others. This reduce the damage resulting from potential cyber risks, for example, loss of critical digital data, reduction of organizations reputation, and other negative effects, and finally minimize cyber risks. In toto, improving cybersecurity measures in accordance with NIST CSF 2.0 core functions reduce success of probable cyber risks. However, this requires comprehensive knowledge of the entities affected, for example, every entity belonging to *Protect* must not only be recognizable, but also continuously monitored and evaluated with regard to potentially anomalous behavior.

Another condition considering the cybersecurity maturity level is the estimation of the effort required to achieve a higher level of cybersecurity maturity. By achieving a higher level of maturity, for example, identified weaknesses (NIS2 Article 21.2.e) in the organizations *Current Profile* of cybersecurity can be remedied. Therefore, it should not be forgotten introducing cybersecurity methods that can usually lead to a change in working situations of employees, which must be positively accompanied by additional, targeted qualification measures (NIS2 Article 21.2.g). Despite this secondary condition, cybersecurity maturity models offer an innovative approach to effectively determining the *Current Profile* in cybersecurity, which is understandable and comprehensible. Below some examples of cybersecurity maturity levels and their ratings are explained:

- A cybersecurity maturity level of *Current Profile* ≤ 3 is assessed in such a way that main obstacles achieving a better level in cybersecurity, have not been eliminated,
- A cybersecurity maturity level of the *Current Profile* $= 4$ is assessed as successful development of capabilities achieving a certain level of sovereignty in cybersecurity,
- A cybersecurity maturity level of the *Current Profile* $= 5$ shows that cybersecurity is firmly anchored in the organization's risk-management strategy, referred as *Target* Profile.

However, the process of cybersecurity maturity effort must focus on the organizations needs due to its business environment. The ability to implement processes from higher maturity level does not imply that maturity levels can be skipped. An organization that is trying to implement a high maturity level (e.g. 5) without the foundation of a defined process (e.g. 4) is likely to fail because of the lack of understanding the impact of process changes.

To summarize, cybersecurity maturity level has become an important instrument for systematically reviewing and gradually improving application-specific capabilities, by means of which different organization-specific dimensions such as strategy, technologies, products, services, and others can be analyzed. This requires an overall picture of the structural concept of representation of capability and maturity levels, discussed in more detail in Sect. 3.3.2.

3.3.2 *Cybersecurity Capability Maturity Models and Classification*

The cybersecurity measures that organizations face is determined by the EU Network and Information Security NIS2 Directive. This include cross-organization and not just organization-wide cybersecurity solutions, for example to comply with

NIS2 Article 21.2.d., and thus overcome existing boundaries within organizations cybersecurity plans. Therefore, cybersecurity Capability Maturity Model (CMM) is a model that illustrates the stages of readiness or preparedness to respond on cyber risks, vulnerabilities and technological advancement that exist within the continuously evolving cyber risk landscape. CMM includes six steps centered on capability-constituents of organizations operational, educational, technical, business as well as legal and regulatory measures. The capability and maturity spectrum include undefined, initial, basic, defined, dynamic and optimizing stages, as described in [43]. CMM relies on the principles of the capability maturity model, like theories of capabilities [44–46], knowledge integration and security practices to guide developments. The CMM that are mainly revealed in scientific research papers [47] are Cybersecurity Capability Maturity Model (CCMM, C2M2), Systems Security Engineering Capability Maturity Model (SSE-CMM), Community Cyber Security Maturity Model (CCSMM) and National Initiative for Cybersecurity Education—Capability Maturity Model (NICE).

Cybersecurity Capability Maturity Model Integration

CMM also include the cybersecurity Capability Maturity Model Integration (CMMI), developed to provide process improvement in organizations. The purpose of CMMI is to provide guidance for improving organization's processes and ability to manage the development, acquisition, and maintenance of products or services. CMMI is noted to aid in improving product quality, reducing cycle time and cost and improving ability to meet project targets. Six capability levels, designated by the numbers 0 through 5, as illustrated in Table 3.34.

CMMI is important for the sectors listed in NIS2 Annex I and II, and describes best practices used in organizations for development, maintenance and procurement of products, services and others within the CMMI framework. This makes it possible to assess the capability and maturity level of organizations in order to put improvements into practice. The CMMI framework contains the distinctive features: [48]:

- **CMMI for Development (CMMI-DEV):** CMMI-DEV is an integrated set of best practices that improves performance and key capabilities for organizations that develop and maintain products, components, and services for digital system development. Benefits of CMMI-DEV are improve time-to-market, increase quality, reduce cost, improve life-cycle management, gain organizational agility. CMMI-DEV replace CMMI for System- und Software Development (CMMI-SE/SW),

CMMI-DEV includes activities for the development and maintenance of products and services that cover the entire life cycle of a product from early conception through market delivery to post-delivery service. Thus, the activities are on development and maintenance of final products. CMMI-DEV is available in two distinct feature variants: CMMI-DEV + IPPD and CMMI-DEV without IPPD. IPPD stands

3.3 Cybersecurity Maturity Models

Table 3.34 CMMI criteria depicted for capability level assignment

Capability level	Capability level assignment	Capability level criteria
0	Incomplete	Undefined process, capabilities primarily non-existent
1	Performed	Process characterized as performed process, characterized by unpredictability, primarily reactive in nature
2	Managed	Characterized by processes that are repeatable. It also uses basic project management to track cost and schedule and is reactive in nature
3	Defined	Proactive process level characterized by defined process that are understood by organization. Standards, procedures, tools, methods are developed to aid completion of tasks
4	Quantitatively Managed	Characterized by measurement where quality and process performance are established and used as criteria in managing the process. Quality and process performance are understood in statistical terms and managed throughout the life of the process
5	Optimizing	Characterized as optimizing process that is changed and adapted to meet relevant current and projected business objectives

for Integration of Product and Process Development. Both variants have a large number of shared activities and are identical in the shared areas.

CMMI-DEV + IPPD contains supplementary goals and practices included in IPPD. The IPPD supplements include an approach whose practices enable organizations achieve close cooperation between the relevant stakeholders throughout the product life cycle in order to serve the needs, expectations and requirements of customers [49]. In order to use these processes that support the IPPD approach, they are combined with other internal organizational processes. If a company uses CMMI-DEV + IPPD-related processes, it has the option of optionally selecting the IPPD additions [48].

Since different maturity models with their characteristic representation types are available for improving specific requirements, a comparison makes it easier to decide which is the most appropriate for the intended application. The two representation types are the representation in capability levels (CMMI) and the representation in maturity levels (Capability Maturity Model - CMM).

The representation in capability levels allows flexible use of CMMI model variants to improve application-related capabilities. Thus, the organization can choose whether only try to improve performance for a single process-related problem, or work on several aspects that are closely linked to the application goal, for example NIS2 Article 21.2.c. In addition, the capability level representation enables different processes to be improved at different speeds. However, dependencies between some process requirements can limit the options. If the processes that require improvement are known, see NIS2 Article 21.2, and the dependencies between the process requirements are clearly described in CMMI, the representation in capability levels is a good choice.

The representation of maturity levels offers a systematic, structured method to approach the model-based process improvement of application goals step-by-step, for example NIS2 Article 21.2.c. Reaching each individual level ensures that appropriate process requirements are laid as a foundation for the next level. Doing so, the process areas are structured according to maturity levels so that the process improvement becomes clearer. The representation according to maturity levels establishes a sequence for implementation of process requirements according to maturity levels, which define the improvement path for the organization from the first, usually suboptimal *Current Profile*, to the optimized level, usually the *Target Profile*. Reaching each individual maturity level ensures that an appropriate improvement basis for the next maturity level is existend, and thus ensures sustainable incremental improvement. If the improvement processes are not directly known or are unclear, see NIS2 Article 21.2, the representation in maturity levels is recommended. These specify a set of processes to be improved at each level, which form the basis for reaching the next maturity level and thus show the maturity path to achieving the goal (*Target Profile*). In this application, the representation in maturity levels is a good choice.

Given that fact, Table 3.35 summarizes both level models in a comparative manner to make it easier to assess which level model is more appropriate for each individual application.

- **CMMI for Services (CMMI-SVC):** CMMI-SVC is an integrated set of best practices that improves performance and key capabilities for organizations that provide services, including Business-2-Business (B2B), a transaction that occurs between two businesses, Business-2-Consumers (B2C), a transaction that takes place between a business and an individual as the end customer, as well as stand-alone services, and those that are part of a product offering. Benefits of CMMI-SVC are gain customer loyalty, develop resilience, improve time-to-market, improve quality, reduce cost. Target group are organizations that provide digital services for other organizations to avoid lack in the ability to react swiftly to cyber risks that impact service delivery, leading to time delays, cost overruns, and failure to meet customer expectations or to over-promise and/or under-deliver.

CMMI-SVC provides guidance for applying CMMI best practices in a service provider organization in the focus on activities providing quality services to customers and end users. CMMI-SVC integrates bodies of knowledge that are essential for a service provider. Thus, CMMI-SVC v1.3 is a collection of service best

Table 3.35 Advantages of level models

Representation in capability levels	Representation in maturity levels
Sequence of improvements can be determined to meet corporate objectives and minimize risks	Predetermined and/or tested improvement path is determined a priori
Better visibility of achieved capabilities for each individual process requirement	Process requirements are defined that are assigned to certain maturity levels
Improving different processes at different speeds	Improvement is indicated in the form of an increased maturity number

practices from government and industry, generated from the CMMI v1.3 Architecture and Framework. CMMI-SVC is based on the CMMI Model Foundation that offers model components common to all CMMI models and constellations, and incorporates work by service organizations to adapt CMMI for use in the service industry. Against this background, CMMI-DEV can be treated as a reference for the development of the service system, which supports delivery of the service.

Given that fact, in CMMI-DEV grades are used to describe a recommended evolutionary path for organizations that want to improve their processes, e.g., for developing their products or services, organization cybersecurity plan, and others. In this regard, CMMI supports two improving options while using levels. One option allows the organization to gradually improve its processes in relation to individual process areas that the organization has previously selected. The other option allows the organization to improve its processes by gradually selecting related and sequential groups of process areas. These two options are based on two types of levels, called capability levels and maturity levels. The representation in capability levels refers to the capabilities within organizations process areas and the representation in maturity levels refers to the overall maturity across a specified set of processes in organizations. This clarification is used for comparisons and appraisals and to guide approaches to improvement. In this context, capability levels refer to how well an organization achieves process improvements in individual process areas. These levels are used to incrementally improve the processes in a given process area. The four capability levels are numbered from 0 to 3. Maturity levels refer to how well an organization achieves process improvements in multiple process areas. These levels are used to improve the processes that belong to a given set of process areas. The five maturity levels are numbered from 0 to 3, or 1 to 5. Against this background, Table 3.36 compares the for four capability and five maturity levels [50].

Notice that the names of two of the levels are the same in both representations (i.e., Managed and Defined) in Table 3.35. The differences are that there is no maturity level 0; there are no capability levels 4 and 5; and at level 1, the names used for capability level 1 and maturity level 1 are different [50].

A capability level is achieved when all of the generic goals are satisfied up to that level. That capability levels 2 and 3 use the same terms as generic goals 2 and 3 is intentional because each of these generic goals and practices reflects the meaning of the capability levels of the goals and practices [46].

Table 3.36 Comparison of capability and maturity levels

Level	Continuous representation capability level	Staged representation maturity levels
Level 0	Incomplete	
Level 1	Informed	Initial
Level 2	Managed	Managed
Level 3	Defined	Informed
Level 4		Qualitatively Managed
Level 5		Optimizing

A short description of each capability level is as follows:

- **Capability Level 0:** Is an incomplete process, not performed or partially performed. One or more specific process goals are not satisfied and no generic goals exist for this level.
- **Capability Level 1:** Informed process accomplishes that the needed work and/or the specific goals of the process area are satisfied. It also results in improvements.
- **Capability Level 2, Managed:** Managed process is planned and executed in accordance with policy; employs skilled people having adequate resources to produce controlled outputs; involves relevant stakeholders; is monitored, controlled, and reviewed; and is evaluated for adherence to its process description.
- **Capability Level 3, Defined:** Defined process is a managed process tailored from the organization's set of standard processes according to the organization's tailoring guidelines; has a maintained process description; and contributes process related experiences to the organizational process assets.

In general, processes advance through capability and maturity levels. In this regard, the maturity level of an organization provides a way to characterize its performance. Hence, a maturity level is a defined evolutionary plateau for organizational process improvement. Each maturity level matures an important subset of the organization's processes, preparing it to move to the next maturity level, measured by the achievement of the specific goals associated with each predefined process. Thus, maturity levels are used to characterize organizational improvement relative to a process, and capability levels characterize organizational improvement relative to an individual process [50].

A short description of each maturity level is as follows:

- **Maturity Level 1, Initial:** Organization usually does not provide a stable environment to support processes. Success in these organizations depends on the competence of the employees in the organization and not on the use of proven processes.
- **Maturity Level 2, Managed:** Projects have ensured that processes are planned and executed in accordance with policy; the projects employ skilled employees who have adequate resources to produce controlled outputs; involve relevant stakeholders; are monitored, controlled, and reviewed; and are evaluated for adherence.
- **Maturity Level 3, Defined:** Processes are well characterized and understood, and are described in standards, procedures, tools, and methods. Organization's set of standard processes, basis for maturity level 3, is established and improved over time, which means that processes are managed more proactively using an understanding of the interrelationships of process activities and detailed measures of the process, its abilities, and its services.
- **Maturity Level 4, Quantitatively Managed:** Organization and projects establish quantitative objectives for quality and process performance and use them as criteria in managing projects. Quantitative objectives help to ensure that subpro-

cess monitoring using statistical and other quantitative techniques is applied to where it has the most overall value to the business.
- **Maturity Level 5, Optimizing:** Organization continually improves its processes based on a quantitative understanding of its business objectives and performance needs. The organization uses a quantitative approach to understand the variation inherent in the process and the causes of process outcomes, and focus on continually improving process performance through incremental and innovative process and technological improvements.

From the previous description of each maturity level, the following rules summarize equivalent staging [48]:

- To achieve maturity level 2, all process areas assigned to maturity level 2 must achieve capability level 2 or 3,
- To achieve maturity level 3, all process areas assigned to maturity levels 2 and 3 must achieve capability level 3,
- To achieve maturity level 4, all process areas assigned to maturity levels 2, 3, and 4 must achieve capability level 3.
- To achieve maturity level 5, all process areas must achieve capability level 3.

Using the staged representation, organizations attain high maturity when achieve maturity level 4 or 5. Achieving maturity level 4 involves implementing all process areas for maturity levels 2, 3, and 4. Likewise, achieving maturity level 5 involves implementing all process areas for maturity levels 2, 3, 4, and 5. Using the continuous representation, organizations attain high maturity using the equivalent staging concept. High maturity that is equivalent to staged maturity level 4 using equivalent staging is attained when organizations achieve capability level 3 for all process areas except for Organizational Performance Management (OPM) and Causal Analysis and Resolution (CAR). High maturity that is equivalent to staged maturity level 5 using equivalent staging is attained when organizations achieve capability level 3 for all process areas.

Cybersecurity SCAMPI Appraisal Method

Oorganizations' often find it useful to measure the progress in cybersecurity using an appraisal to obtain a capability profile or a maturity rating. Appraisals are conducted for one or more reasons:

- Determine the extent to which the organization's processes meet CMMI best practices and areas for improvement,
- Keep customers and suppliers informed about the extent to which organization's processes meet CMMI best practices,
- Meet contractual requirements of customers.

Appraisals by organizations using a CMMI model to meet the requirements set out in the Appraisal Requirements for CMMI (ARC) document. ARC requirements are designed to help improve appraisal consistency across multiple constellations, models, and appraisal methods to understand the tradeoffs associated with various methods. Moreover, appraisals serve the purpose of identifying opportunities for improvement and comparing organization's processes with CMMI best practices. Hence, Appraisal teams use a CMMI model and an Appraisal Requirements for CMMI (ARC) method to guide in assessing the organization and presenting the results, which then are used to plan further improvements for the organization.

A particular appraisal method is declared as an ARC Class A, B, or C appraisal method, depending on which set of ARC requirements the method developer considered when designing the method. More information about ARC can be found on the SEI website [51].

The Standard CMMI Appraisal Method for Process Improvement (SCAMPI) is the official Software Engineering Institute (SEI) method to provide benchmark-quality ratings relative to CMMI models to evaluate organizations' processes and provide ratings. SCAMPI includes three appraisal classes. SCAMPI A appraisals are officially recognized appraisals that result in benchmark quality ratings (e.g., maturity levels). SCAMPI B & C appraisals are less rigorous appraisals designed to provide information for process improvement or the status of process improvement work SCAMPI assessment methods used for conducting assessments using CMMI model variants. The SCAMPI Method Definition Document (SCAMPI-MDD) contains rules for ensuring the consistency of assessments. In order to enable comparison with other companies, for example, uniform assessments must be guaranteed. Therefore, achieving a certain level or fulfilling a required application behavior, e.g. cybersecurity according to NIS2 Article 21.2.d., must have the same meaning for different companies in the supply chain. For this purpose, the SCAMPI assessment method family includes classes A, B and C [52].

- **SCAMPI A:** Most rigorous method and the only method that result in a rating as benchmarks to compare maturity or capability levels across organizations.
- **SCAMPI B:** Appraisals are recommended for initial assessments in organizations that are beginning to use CMMI models for process improvement activities.
- **SCAMPI C:** Appraisals most likely used when the need for a *quick view* arises or for periodic self-assessments by project and organizational support groups. It provides a wide range of options, including characterization of planned approaches to process implementation according to a scale defined by the user.

When using SCAMPI, the following boundary conditions apply to a CMMI-based assessment system [48]:

- Selection of CMMI model variant for assessment, i.e. CMMI-DEV or CMMI-DEV + IPPD,
- Determination of the scope of the assessment, e.g. business unit under consideration, CMMI application areas to be examined (according to NIS2 Article 21.2), and capability or maturity levels to be assessed,
- Selection of assessment method,
- Selection of assessment team,

- Selection of assessment participants to be surveyed
- Allocation of assessment results, e.g. classifications or instantiation of specific results,
- Determination of assessment restrictions, e.g. time required (on site). SCAMPI-MDD enables the selection of predefined options for an assessment. The assessment options help companies to align the CMMI model variant with the requirements and objectives, e.g. NIS2 Article 12.2.

What outcome can be expected from SCAMPI A [52]:

- Capability level ratings or maturity level rating.
- Findings that describe the strengths and weaknesses of organization's process relative to the CMMI.
- Consensus regarding the organizations' key process issues.
- An appraisal database that the organization can continue to use to monitor process improvement progress and to support future appraisals.

What outcome can be expected from SCAMPI B [52]:

- Detailed findings that describe the strengths and weaknesses of organizations' process relative to the CMMI.
- Practice characterizations indicating the likelihood that the examined practices would satisfy the goals and meet the intent of the CMMI.
- Consensus regarding the organizations' key process issues.
- A FIDO Database that the organization can continue to use to monitor process improvement progress and support future appraisals. FIDO (Fast IDentity Online) is a set of open, standardized authentication protocols intended to ultimately eliminate the use of passwords for authentication.

What outcome can be expected from SCAMPI C [52]:

- Describe the strengths and weaknesses of the assessed processes.
- Depending on the appraisal scope and strategy, findings may be mapped to the relevant CMMI components.
- Summarize the adequacy of the assessed processes vis-á-vis the CMMI.
- Process improvement actions.
- A FIDO Database that the organization can continue to use to monitor process improvement progress and to support future appraisals.

3.4 Cybersecurity Maturity Assessment

Conducting a cybersecurity maturity assessment is essential for organizations of all sizes and sectors to evaluate their cybersecurity posture and identify potential areas of improvement. A thorough assessment enables measure organizations ability to detect, prevent, and respond to cyber risks sustainable. Hence, a cybersecurity maturity assessment begins with clearly defined objectives and scopes, by means to determine the specific goals to achieve the assessment, such as improving overall cybersecurity, identifying vulnerabilities, or meeting compliance requirements such as NIS2. This requires to consider critical assets, systems, and processes that should be evaluated, including internal and external systems, cloud services, third-party dependencies, and others. A well-defined scope ensures that the assessment remains focused and delivers actionable insights. From end to end, a multi-level approach is required as described in [53], followed in the subsequent bold titles.

Gather Information and Conduct Interviews: Application-oriented data collection and detailed insights from key stakeholders is a key issue of a comprehensive cybersecurity maturity assessment. The collected data are related to cybersecurity policies, procedures, and existing cybersecurity controls. Encourage interviews with key personal in the organization, such as IT personnel, cybersecurity teams, business unit leaders, and others acquire valuable insights into current cybersecurity practices and potential areas for improvement.

Assess Risk-Management: Evaluate organization's current risk-management practices to determine the ability to identify and mitigate cyber risks effectively. Capitalize current risk assessments, risk prioritization, and risk mitigation plans and ensure that identified cyber risks are appropriately categorized to their potential impact and that suitable reduction measures are in place.

Analyze Incident Response Capabilities: Analyze incident response capabilities that play crucial roles in organizations, mitigating the impact of cyber risks. Thus, organizations should evaluate identification, containment, eradication, and recovery procedures in the sense to a potential cybersecurity breach. Review current incident response plan for alignment with sector specific best practices and organization's specific needs. Effective incident response minimizes the impact of cyber threat risk incidents and ensures swiftly return to normal operations.

Evaluate Access Controls: Examine current access controls and authentication mechanisms to verify that only authorized personnel can access sensitive information and critical assets. Assess the implementation of Multi-Factor Authentication (MFA) and Privileged Access Management (PAM) to safeguard against unauthorized access and potential data breaches. MFA is a multi-step account login process that requires users to enter more information than just a password. PAM is an identity cybersecurity solution that describe the policies and technologies used to protect, control, and audit access to privileged accounts, credentials, and resources in organization's IT infrastructure.

Measure Data Protection Measures: Data protection review is a critical aspect of a cybersecurity maturity assessment. Evaluate current data protection measures, data masking, and classification to safeguard sensitive information from unauthorized access. Assess the organization's digital data protection practices to ensure their CIA triad (see Sect. 1.5).

Review Security Awareness and Training: Human factor is a significant component of cybersecurity posture. Assess the current effectiveness of Cybersecurity Awareness (CSA) and cybersecurity training programs for employees to gauge their understanding of cybersecurity policies and procedures. An educated and cybersecurity-conscious workforce serve as an indispensable defense line against social engineering attacks and other cyber risks.

Benchmark Against Industry Standards: Cybersecurity frameworks enable a comprehensive perspective on organization's cybersecurity maturity, benchmark it against international cybersecurity standards such as NIST CSF 2.0, ISI/IEC 27K, CIS Controls, MITRE ATT&CK, NIS2, and others. These standards offer valuable guidelines and benchmarks enabling to identify areas for improvement and prioritize cybersecurity initiatives.

Score and Report Findings: After completing the assessment and understanding organization's cybersecurity capabilities, assign scores to different domains based on evaluation results. Develop a report that outlines a comprehensive view of strengths, weaknesses, and recommended actions to enhance a cybersecurity program. A well-structured report communicates assessment findings clearly to key stakeholders and serves as basis for formulating improvement strategies [53].

Develop a Security Posture Improvement Plan: Based on the assessment findings and recommendations, an actionable improvement plan must be created. The plan should outline specific steps, timelines, and responsible stakeholders for addressing identified weaknesses and enhancing cybersecurity maturity. Prioritize critical areas for improvement and allocate resources effectively to implement the requiredchanges [53].

A thorough cybersecurity maturity assessment provides organizations with valuable insights into their cybersecurity posture and enables to identify strengths as well as vulnerabilities. By following the detailed steps above, organizations can effectively evaluate risk-management, incident response, access controls, data protection, and employee CSA.

With proper planning, evaluation, and understanding, organizations can make informed decisions to strengthen their cybersecurity defenses and stay resilient in the face of ever-evolving cyber threat risk incidents. Substantial is a proactive approach taking a cybersecurity maturity assessment as key to safeguarde valuable digital assets in an increasingly complex digital environment.

3.5 Exercises

- What is meant by the term *Risk-Management*?
    - Describe the characteristics and capabilities of Risk-Management
- What is meant by the term *Risk Identification and Documentation?*
    - Describe the characteristics and capabilities of Risk Identification and Documentation
- What is meant by the term *Risk Quantification and Documentation?*
    - Describe the characteristics and capabilities of Risk Quantification and Documentation

- What is meant by the term *Assessment and Documentation?*
  - Describe the characteristics and capabilities of Risk Assessment and Documentation
- What is meant by the term *Contingency Planning?*
  - Describe the characteristics and capabilities of Contingency Planning
- What is meant by the term *NIST CSF 2.0?*
  - Describe the characteristics and capabilities of NIST CSF2.0
- What is meant by the term *NIST CSF 2.0 Core Functions?*
  - Describe the characteristics and capabilities of the six NIST CSF 2.0 Core Functions
- What is meant by the term *NIST CSF 2.0 Profile?*
  - Describe the characteristics and capabilities of NIST CSF2.0 Profile
- What is meant by the term *NIST CSF 2.0 Tiers?*
  - Describe the characteristics and capabilities of NIST CSF2.Tiers
- What are the *Benefits of Adopting NIS CSF2.0?*
  - Describe at the meaning of the five key benefits
- What is meant by the term *MITRE ATT&CK?*
  - Describe the characteristics and capabilities of MITRE ATT&CK
- What is meant by the term *MITRE Engage?*
  - Describe the characteristics and capabilities of MITRE Engage
- What is meant by the term *CISCSC?*
  - Describe the characteristics and capabilities of CISCSC
- What is meant by the term *ISO/IEC27K?*
  - Describe the characteristics and capabilities of ISO/IEC 27K
- What is meant by the term *Cybersecurity Maturity Level Model?*
  - Describe the characteristics and capabilities of Cybersecurity Maturity Level Models
- What is meant by the term *Cybersecurity Capability Maturity Model?*
  - Describe the characteristics and capabilities of Cybersecurity Capability Maturity Models.

References

1. Directive (EU) 2022/2555 of the European Parliament and of the Council of 14 December 2022 on measures for a high common level of cybersecurity across the Union, amending Regulation (EU) No 910/2014 and Directive (EU) 2018/1972, and repealing Directive (EU) 2016/1148

References

1. (NIS 2). https://eur-lex.europa.eu/legal-content/EN/TXT/PDF/?uri=CELEX:32022L2555. (accessed 01.2023)
2. Regulation (EU) 2016/679 of the European Parliaments and of the Council of 27 April 2016, on the protection of natural persons with regard to the processing of personal data and on the free movement of such data, and repealing Directive 95/46/EC (General Data Protection Regulation). ELI: http://data.europa.eu/eli/reg/2016/679/oj. (accessed 01.2023)
3. Framework for Improving Infrastructure Cybersecurity, Version 1.1. https://nvlpubs.nist.gov/nistpubs/CSWP/NIST.CSWP.04162018.pdf. (accessed 01.2022)
4. MITRE ATTACK (MITRE ATT&CL) Framework. https://attack.mitre.org. (accessed 02.2022)
5. CIS Critical Security Controls. https://learn.cisecurity.org/cis-controls-download
6. ISO/IEC 27K. https://www.iso.org/obp/ui/en/#iso:std:73906:en. (accessed 03.2023)
7. Möller, D.P.F.: Guide to Cybersecurity in Digital Transformation: Trends. Methods, Technologies, Applications and Best Practices
8. Risk assessments: What they are, why they are important and how to complete them. 2024. https://www.britsafe.org/training-and-learning/informational-resources/risk-assessments-what-they-are-why-they-re-important-and-how-to-complete-them. (accessed 07.2024)
9. What is risk identification? Definition and Tools. 2024. https://www.indeed.com/career-advice/career-development/risk-identification. (accessed 10.2024)
10. Risk Management Framework for Information Systems and Organizations - A System Life Cycle Approach for Security and Privacy. 2018. https://nvlpubs.nist.gov/nistpubs/SpecialPublications/NIST.SP.800-37r2.pdf. (accessed 10.2022)
11. ISO/IEC 27005:2022 Information security, cybersecurity and privacy protection — Guidance on managing information security risks. https://www.iso.org/standard/80585.html. (accessed 07.2022)
12. Zola, A.: Risk Assessment Framework (RAF). 2022. https://www.techtarget.com/searchcio/definition/risk-assessment-framework-RAF. (accessed 07.2022)
13. Siddiqui, M.: A Pocket Guide to Factor Analysis and Information Risk (FAIR). 2024. https://www.cybersaint.io/blog/a-pocket-guide-to-factor-analysis-of-information-risk-fair. (accessed 06.2024)
14. https://www.coso.org/guidance-on-ic. (accessed 07.2023)
15. A Business Framework for the Governance and Management of Enterprise IT. 2012. https://files.santaclaracounty.gov/migrated/COBIT-5_res_eng_1012%20%28ISACA%29.pdf. (accessed 05.2023)
16. Alberts, A. J., Behrens, S. G., Pethia, R. D., Wilson, W.R.: Technical Report CMU/SEI-99-TR-017 ESC-TR-99-017, OCTAVE, 1999. https://insights.sei.cmu.edu/documents/1210/1999_005_001_16769.pdf. (accessed 05.2023)
17. Wynn, J.: Threat Assessment and Remediation Analysis (TARA). 2014. https://www.mitre.org/sites/default/files/2021-10/pr-14-2359-tara-introduction-and-overview.pdf. (accessed 05.2022)
18. The NIST Cybersecurity Framework (CSF) 2.0. 2024. https://nvlpubs.nist.gov/nistpubs/CSWP/NIST.CSWP.29.pdf. (accessed 05.2024)
19. NIST Releases Version 2.0 of Landmark Cybersecurity Framework. 2024. https://www.nist.gov/news-events/news/2024/02/nist-releases-version-20-landmark-cybersecurity-framework. (accessed 05.2024)
20. Strom, B. E., Applebaum, A., Miller, D. P., Nickels, K. C., Pennington, A. G., Thomas, C. B.: MITRE ATT&CK ®: Design and Philosophy. 2020. https://attack.mitre.org/docs/ATTACK_Design_and_Philosophy_March_2020.pdf. (accessed 05.2022)
21. Strom, B. E., Battaglia, J. A., Kemmerer, M. S., Kupersanin, W., Miller, D. P., Wampler, C., Whitley, S. M., Wolf, R. D.: Finding Cyber Threats with ATT&CK™-Based Analytics. 2017. https://www.mitre.org/sites/default/files/2023-08/16-3713-finding-cyber-threats-with-attack-based-analytics.pdf. (accessed 05.2022)
22. MITRE Engage. https://engage.mitre.org. (accessed 03.2025)

23. MITRE Engage – A Practical Guide to Adversary Engagement, 2022. https://engage.mitre.org/wp-content/uploads/2022/03/EngageHandbook.pdf. (accessed 03.2025)
24. What is Deception Technology? Zscaler, 2025. https://www.zscaler.com/resources/security-terms-glossary/what-is-deception-technology. (accessed 03.2025
25. Heckmann, K. E., Stech, F. J., Thomas, R. K., Smoker, B., Tsow, A. W.: Cyber Denial, Deception and Counter Deception. A Framework for Supporting Active Cyber Defense. Springer International Publishing, 2015
26. CIS Critical Security Controls Version 8.1. https://www.cisecurity.org/controls/v8-1. (accessed 05.2024)
27. A Roadmap to the CIS Critical Security Controls. 2024. https://www.cisecurity.org/controls/v8-1. (accessed 05.2024)
28. Guide to Implementation Groups (IG) CIS Critical Security Controls v8.1. 2024. https://learn.cisecurity.org/controls-v8.1-guide-to-implementation-groups. (accessed 05.2024)
29. The Ongoing Evolution of the CIS Critical Security Controls. https://www.cisecurity.org/insights/blog/ongoing-evolution-cis-critical-security-controls. (accessed 05.2024)
30. Boeckman, A.,Ferguson, B., Klingbile, K., Opatrny, J., Regnier, R. (CIS), Youngm J. (CIS) (Eds..),Carte, B. (CIS). Franklin, J. (CIS), Wicks, M. (CIS) (Contributors): CIS Security Controls® v8.1 Industrial Control Systems (ICS) Guide. 2024. https://learn.cisecurity.org/cis-controls-v8.1-ics-guide. (accessed 05.2022)
31. https://stock.adobe.com/de/search?k=spider%20chart. (accessed 05.2022)
32. Weckenmann, A., Akkasoglu, G.: Maturity Determination and Information Visualization of New Forming Processes considering Uncertain Indicator Values. In: American Institute of Physics Conference Proceedings, Vol. 1431, pp. 899 ff, 2012. https://doi.org/10.1063/1.4707649
33. von Altrock, C., Krause, B.: Multi-criteria Decision Making in German Automotive Industry Using Fuzzy Logic. Fuzzy Sets and Systems, Vol. 63, No. 3, pp. 375-380, 1994
34. Akkasoglu, G.: Methodology for Conception and Application-specific Maturity Models considering Information Uncertainty (in German). PhD Thesis, University of Erlangen Nuremberg, 2013
35. Möller, D., Iffländer, L., Nord, M., Leppla, B., Krause, P., Czerkewsky, P., Lenski, N., Mühl, K.: Cybersecurity in the Railway Sector. In: Proceedings 17th International Conference on Critical Information Infrastructures Security. In: LNCS, Springer Publ., 2022
36. Möller, D.P.F., Iffländer, L., Nord, M., Krause, P., Leppla, B., Mühl, K., Lenski, N., Czerkewski, P.: Emerging Technologies in the Era of Digital Transformation: State of the Art in the Railway Sector. In: Proceed. 19th Int. Conference on Informatics in Control, Automation and Robotics, pp. 721-726. Cite Press 2022
37. Becker J, Knackstedt D, Pöppelbuß J.: Developing Maturity Models for IT Management – A Procedure Model and its Application. In: Business Information Systems Engineering, 213–222, 2009
38. Dube, D. P., Mohantly, R. P.: Towards Development of a Cybersecurity Capability Maturity Model. Intern. J. Business Information Systems, 34, 1, 104–127, 2020
39. Jeston J, Nelis C.: Business Process Management: Proacted Guidelines to Successful Implementation. Elsevier Publ., 2201
40. Venkatraman V.: The Digital Matrix: New Rule for Business Transformation through Technology. Life Tree Book Publ. 2017
41. Rogers D. I.: The Digital Transformation Playbook: Rethink your Business for the Digital Age. Columbia University Press, 2006
42. Möller D.P.F., Vakilzadian H.: Cybersecurity Risk in Digitalization of Infrastructure Systems: A Use Case. In: Proceedings IEEE-EIC International Conference, 2023. https://doi.org/10.1109/eIT57321.2023.10187348. (accessed 10.2023)
43. Barclay, C.: Sustainable Security Advantage in a changing Environment: The Cybersecurity Capability Maturity Model. 2014. https://www.researchgate.net/publication/262917234_Sustainable_Security_Advantage_in_a_Changing_Environment_The_Cybersecurity_Maturity_Model. (accessed 01.2022)

References

44. Teece, D. J., Pisano, G., Shuen, A.: Dynamic Capabilities and Strategic Management. Strategic Management Journal, 18, 7, 509–533, 1997
45. Teece, D. J., Pisano, G.: The Dynamic Capabilities of Firms an Introduction. Industrial and Corporate Change, 3, 3, 537–556, 1994
46. Grant, R. M., Prospering in Dynamically-Competitive Environments: Organizational Capability as Knowledge Integration. Organization Science, Vil. 7, No. 4, pp. 375–387, 1996
47. Mohammed, I., Bade, A.M.: Cybersecurity Capability Maturity Model for Network System. International Journal of Development Research, Vol. 9, No. 7, pp. 28637–28641, 2019
48. Chrissis, M. B., Konrad, M., Shrum, S.: CMMI® - Guidelines for Process Integration and Product Improvements (in German). Addison Wesley Publ., 2006
49. DoD Guide to Integrated Product and Process Development (Version 1.0). Washington, DC: Office of the Under Secretary of Defense (Acquisition and Technology), 1996. www.abm.rda.hq.navy.mil/navyaos/content/dow-115] oad/1000/4448/file/ippdhdbk.pdf. (accessed 01.2022)
50. CMMI® for Development, Version 1.3. Technical Report, Reference to CMU/SEI-2010-TR-033, Reference to ESC-TR-2010-033. Carnegie Mellon University. 2010
51. Appraisal Requirements for CMMI (ARC). www.sei.cmu.edu/cmmi/appraisals/appraisals.html. (accessed 01.2022)
52. SCAMPI: Standard CMMI Appraisal Method for Process Improvement. https://www.plays-in-business.com/scampi-standard-cmmi-appraisal-method-for-process-improvement/. (accessed 01.2022)
53. Mignacca, J.: Cybersecurity Maturity Assessment. 2023. https://www.cavelo.com/blog/cybersecurity-maturity-assessment (accessed 08.2023)

Chapter 4
Application Domain Network and Information Security (NIS2)

4.1 Network and Information Security (NIS2)

With an escalating array of cyber-attack surfaces and vectors as a result of the current comprehensive digitalization of all areas of life, and the ubiquitous connecting in digital transformation in every facet of life and economic business, cybersecurity is still paramount and need to strengthen cyber risk defenses. Therefore, government organization lay down cybersecurity directives like the European Union (EU) Directive on the Resilience of Critical Entities (RCE), which include obligations on the 27 EU Member States to take dedicate measures, to ensure that essential provisions and services for the maintenance of vital societal functions or economic activities are provided in an unobstructed manner in the EU internal market. The RCE Directive build a framework to support the EU-Member States in ensuring that critical entities have the capability to prevent, resist, absorb and recover from disruptive incidents, including those caused by natural hazards, terrorism, insider threats, sabotage, or public health emergencies. In this regard, the RCE and the new NIS2 Directive define the frameworks for EU-Member States critical infrastructure protection. Both RCE (EU 2022/) [1] and NIS2 (EU 2022/2555) [2] were enacted at the end of 2022 in the EU and need to be transposed into national law in the EU Member States.

In this context some key objectives of NIS2 include:

- Expanding the scope of cybersecurity to cover more business sectors vital to the EU-Member States,
- Mandating a proactive stance in regard to stricter and more comprehensive cybersecurity risk-management,
- Set stricter reporting obligations for entities to report significant cyber threat risk incidents that could cause severe operational disruption, and require enhancing monitoring and detection,

- Fostering on securing the entire supply chain, including third-party vendors and service providers,
- Fostering greater cooperation between EU-Member States,
- Assigning regulatory authority to competent national authorities,
- Allowing penalties for non-compliance,
- And others.

Therefore, the EU enhanced the former initiated Network and Information Security (NIS) Directive into the new NIS2 Directive to uniform regulate cybersecurity objectives, and to achieve a high common level of cybersecurity across the EU Member States due to the growing demand on cybersecurity required by digital transformations interconnectedness of essential digital systems and devices, infrastructure resources and others. The development to NIS2 as new cybersecurity directive with a focus on RCE, the network of Security Operations Centers (SOCs) and expanded measures to strengthen the cyber resilience, because the cybersecurity risk landscape has changed considerably. Against this background, the NIS2 Directive refers to:

- Reinforced rules for a high common cybersecurity level across the EU Member States,
- Supervision for medium and large organizations across sectors of high criticality (NIS2 Annex I, 11 Sectors) and other critical Sectors (NIS2 Annex II, 7 Sectors),
- Establishment of higher levels of mandatory, reviewable and sanctionable cybersecurity measures for risk-management aligned with standards like ISO/IEC 27 K, MITRE ATT&CK, NIST CSF 2.0 and others, cybersecurity governance, incident reporting and recovery, resilience, network, and information system and application cybersecurity,
- Cybersecurity risk reviews of cybersecurity practices for major connected third-party services providers,
- Conduct regular vulnerability assessments and remediation,
- Perform Multi-Factor Authentication (MFA) and stringent password policies,
- Deploy endpoint protection through antivirus and Endpoint Detection and Replace Solution (EDR) that record and store endpoint-system-level behaviors, use various data analytics techniques to detect suspicious system behavior, provide contextual information, block malicious activity, and provide remediation suggestions to restore affected systems. EDR solutions must provide the following four primary capabilities:
    - Detect cyber risks,
    - Contain the cyber risk at the endpoint,
    - Investigate cyber risk incidents,
    - Provide remediation guidance,

NIS2 correlates with the RCE Directive and address the hybrid cyber-physical Operational Technology (OT). OT represent a broad range of programmable systems and devices that interact with the physical environment to control and/or

4.1 Network and Information Security (NIS2)

manage systems and devices. OT also detect or cause an influence of the behavior through monitoring and/or control of devices, processes and others, often in real-time.

The RCE Directive is trans posited in parallel to the NIS2 Directive. Both address current and future online and offline cyber risks from cyber threat risk incidents and/or natural disasters. RCE focuses on physical rather than digital resilience measures for critical entities. Therefore, national authorities perform reviews, and assess the effectiveness and accountability of the risk-management process, and provide significant support to their critical and crucial entities. Currently, the NIS2 Directive be official in force by EU parliament vote. Thereafter, the adoption by national legislative bodies across the EU-Member States is laid down, so that NIS2 can come into effect October 17th, 2024 or on the latest in 2025.

The European Union Agency for Cybersecurity (ENISA) is the EU's agency dedicated to achieve a high common level of cybersecurity across EU-Member States, has published in 2020 a report about NIS [3], which presents the findings of a survey, examining approaches to cybersecurity of some EU-Member States. Compared to their US counterparts, data shows that the EU organizations allocate on average 41% less to cybersecurity than their US counterparts [4]. This is a worrying situation given the fact of availability of increasingly sophisticated spy software. Spyware is software with malicious behavior that aims to gather information to harm a person and/or organization.

Pegasus is such a spyware developed by the Israeli cyber-arms company NSO Group, named after Pegasus a winged horse of the Greek mythology. Pegasus is a Trojan horse computer virus. As of 2022 Pegasus was capable of reading text messages, tracking calls, collecting passwords, location tracking, accessing the targeted devices microphone and camera, and harvesting information from in-stalled apps [5–8]. Pegasus being widely used against high-profile and critical targets. News of the Pegasus spyware caused significant media coverage that called the most sophisticated smartphone attack (iOS and Android) ever. It was the first time that a malicious remote exploit used iOS jailbreaking to gain unrestricted access to an iPhone [5]. Jailbraking is use of a privilege escalation exploit to remove software restrictions imposed by the software manufacturer.

Another commercial spyware capable of infiltrating mobile phones and stealing everything digital inside of them like videos, pictures, text messages, search history, passwords, call logs and other is Predator. Predator spyware typically sold to high-paying government clients by a company called Cytrox, a secretive surveillance firm based in North Macedonia. Cytrox is owned by an Israeli parent company called Intellexa [9]. Furthermore, the New York Times obtained a copy of a nine-page Intellexa pitch for Predator to a Ukrainian intelligence agency in 2021, the first full such commercial spyware proposal to be made public [10].

Oqing the rapid development of increasingly complex and sophisticated malware, the Executive Office of the President of the United States already has published the Cybersecurity Enhancement Act (Public Law 113–274) in 2014, which requires the National Science and Technology Council and the Networking and Information Technology Research and Development Program to develop, maintain, and update every 4 years a Cybersecurity Research and Development Strategic Plan. This plan also addresses priorities established by the 2018 National Cyber Strategy of the United States, including both its domestic and foreign priorities, and by the respective Administrations FY Research and Development Budget Priorities Memorandum. The Plan identifies the following goals for cybersecurity research and development [11]:

- Understand human aspects of cybersecurity,
- Provide effective and efficient risk-management,
- Develop effective and efficient methods for deterring and countering malicious cyber threat risk incident activities,
- Develop integrated safety-security framework and methodologies,
- Improve systems development and operation for sustainable cybersecurity.

To realize the goal of a cybersecure cyberspace, the plan carries forward the essential concepts from the 2016 Federal Cybersecurity Research and Development Strategic Plan [12], including the framework of four interdependent defensive capabilities [11]:

- **Deter:** Ability to discourage malicious cyber threat risk incident activities by increasing the cost to, diminishing the spoils of, and increasing the risks and uncertainty of potential cyber-attackers,
- **Protect:** Ability of components, systems, users, and critical Infrastructure to efficiently resist malicious cyber threat risk incident activities, and to ensure the CIA Triad (see Sect. 1.5) and accountability,
- **Detect:** Ability to efficiently detect, and even anticipate, cyber-attacker decisions and activities, given that perfect cybersecurity is not possible and that computer systems, networks and information systems, infrastructure resources and other should be assumed to be vulnerable to malicious cyber-threat risk incident activities,
- **Respond:** Ability of defenders, defenses, and infrastructure to react dynamically to malicious cyber threat risk incident activities by efficiently adapting to disruption, countering the malicious activities, recovering from damage, maintaining operations while completing restoration, and adjusting to that similar future activities.

These four elements are similar but not identical to the six core functions in the National Institute of Standards and Technology's (NIST) Cybersecurity Framework (CSF) 2.0 for improving cybersecurity in critical infrastructures and digital systems.

To advance the priorities and objectives of the 2018 National Cyber Strategy of the United States and the Administration's FY Research and Development Budget Priorities Memorandum, the Plan outlines research objectives in the following priority areas [13]:

- **Artificial Intelligence (AI):** Analyze and correlate event and cyber threat risk incident data across multiple sources, turning it into clear and actionable insights that cybersecurity analysts use for further investigation and defending, as well as cyber threat risk incident attack mitigation.

- **Quantum Information Science (QIS):** Based on algorithms that perform computations faster than known in classical algorithm. Thus, QIS is expected to become important in detecting cyber threat risk incident attacks in the early stages before any significant damage occur. It also is assumed to be used to develop robust cryptography standards to provide significant protections for digital data.
- **Trustworthy Distributed Digital Infrastructure:** Technologies that facilitate cybersecure information communications infrastructure that enables next-generation wireless communication, distributed computing, seamless integration of telecommunication systems with cyber-physical systems, and provides the communications infrastructure for the Industries of the Future (IotF) [13].
- **Privacy:** Assurance that the confidentiality of, and access to, certain information about an entity is protected.
- **Secure Hardware and Software:** Hardware cybersecurity involves protecting devices from tampering and unauthorized access. Hardware encryption, for example, cybersecure data by encoding it at a chip level. This protection is crucial because it safeguards sensitive information before potential cyber threat risk incident attackers can reach software defenses. Software cybersecurity protects data and programs from malicious harm, defending potential cyber threat risk incident attackers from using software flaws to break into software systems. Regular updates fix these issues, keeping software cyber safe.
- **Education and Workforce Development:** Involves equipping employees with the knowledge, skills, and expertise necessary to effectively identify and mitigate cyber threat risk incidents.

Advancements in the defensive capabilities depend on progress in human knowledge, skills, and expertise, organizations research infrastructure, risk-management, scientific foundations, and the transition to practice [11].

Thus, enhancing the trustworthiness of networks and information systems, infrastructure resources and others is still paramount, which requires enhancing developing and deploying practices of cybersecurity strategies. Promising research involves integration of AI and Machine Learning (ML) methods to detect program errors, identify vulnerabilities, and make it easier for software engineers to develop Security by Design and Default (see Sect. 1.3.5) within their products/systems.

Against the background of the above facts, substantial activities are required to be prepared for the NIS2 requirements, which the following subchapters discuss in detail.

4.2 Compliance and Regulatory Pressure

As can be seen from NIS2 Chapter II Coordinated Cybersecurity Frameworks and Chapter III Cooperation at Union and International Level that organizations must take appropriate and proportionate technical and organizational measures to ensure cyber resilience. In this context:

- Occurrence of cyber threat risk incidents must be prevented,

- Appropriate protection of sensitive areas, facilities and infrastructure must be ensured, including access controls,
- Prevention of cyber threat risk incidents and their possible consequences must be limited,
- Securing networks and information systems by effective implementing cybersecurity policies,
- Implementation of risk and crisis management procedures and protocols and specified chains of action in the event of an alarm must be established,
- And other.

After potential cyber threat risk incidents, operational business processes must be restored immediately by measures taken to maintain them and, if necessary, alternative supply chains identified. The following key indicators are indispensable: Recovery Time Objective (RTO), Recovery Point Objective (RPO), Maximum tolerable Downtime (MtD). Beside this, appropriate management to ensure employee safety is needed, for example by:

- Defining categories for personnel who perform critical functions,
- Defining access rights to sensitive areas, facilities, infrastructure and sensitive information,
- Raising staff cybersecurity awareness.

Furthermore, managing directors must implement NIS2 risk-management measures and monitor their implementation (NIS2 Article 18 Report on the state of cybersecurity in the EU) by technical and organizational measures to ensure a high cybersecurity level. In addition, managing directors can be held personally liable for non-compliance, as described, for example, in NIS2 Implementation Act—NIS2 UmsuCG in Germany—Chapter 2 Risk-Management, Reporting, Registration, Proof and Information Obligations, Article 38 Approval, monitoring and training obligation for management of particularly important facilities and important facilities [14]. In this regard fines of up to €10 million or 2% of global turnover are introduced, for example, NIS2 Implementation Act—NIS2 UmsuCG—Chapter 7 Sanctions and Supervision, Article 60 Fine provisions. NIS2 UmsuCG is the German implementation of the NIS2 Directive to control the information security management in the German Government Administration. Thus, ensuring network and information system cybersecurity is essential, because supervisory authorities can act up to and including suspending business operations, for example, NIS2 Implementation Act—NIS2 UmsuCG—Chapter 7 Sanctions and Supervision, Article 64 Central competence in the EU for certain types of entities. Against this fact, members of management bodies are obliged to attend cybersecurity training to clearly understand their potential impact on the organization. This ensures that the necessary knowledge and skills to effectively assess and (proactively) respond to potential cyber threat risk incidents and their impact on business activities is achieved.

The supervision of high critically sectors (NIS2 Annex I Sectors of High Criticality) and other critical sectors (NIS2 Annex II Other Critical Sectors) by

4.2 Compliance and Regulatory Pressure

national authorities is regulated by law. EU-Member States must ensure that competent authorities can carry out the following measures:

- Inspections inside and outside the respective sector organization, including randomized checks,
- Regular audits (high critically sectors),
- Cybersecurity checks based on risk assessments or available cyber-risk related information,
- Cybersecurity scans based on objective, fair, transparent and non-discriminatory risk assessment criteria,
- Information required to assess the sectors facility's cybersecurity measures, including documentation of cybersecurity strategies and compliance with the reporting obligation to ENISA (NIS2 Article 16 EU cyber crisis liaison organization network (EC-CyCLONe) and Article 20 Governance),
- Requests for access to data, documents and information required to fulfil supervisory tasks,
- Documentation of the implementation of cybersecurity strategies, e.g., results of cybersecurity audits (conducted by certified auditors) and underlying documents (for high critical sectors),
- Issuing alerts in the event of non-compliance with the obligations as set out in NIS2,
- Issuing binding instructions or requesting public authorities to promptly remedy deficiencies or violations of obligations as set out in NIS2,
- Requesting behavior that violates NIS2 obligations be stopped and refrained from repeating,
- Requesting risk-management measures and reporting obligations be brought into line with the obligations in NIS2 Article 18 Report on the state of cybersecurity in the EU, and Article 20 Governance within a specified time period,
- Requiring persons or organizations that provide services or activities to an organization that may be affected by a significant cyber threat risk incident to inform about all measures and/or actions that may be necessary to avert a potential cyber threat risk incident,
- Issuing a public statement naming persons/organizations teams responsible for violations of obligations as set out in NIS2 and describing the nature of the violation,
- Instructing competent authorities or courts to impose a fine in accordance with national law and/or NIS2, Article 31 General aspects concerning supervision and enforcement, considering the circumstances of the individual case.

In the case of serious violations and the criticality of the provisions violated, sanctions will be imposed. In this context, reference will be made to a set of facts to classify serious violations and/or repeated violations:

- Failure to report or resolve cyber threat risk incidents with a significant disruptive effect,
- Failure to remedy deficiencies despite binding instructions from competent authorities,

- Hammering ordered audits or monitoring activities after a violation has been identified, providing inaccurate or clearly inaccurate information in relation to risk-management requirements or reporting obligations under NIS2 Article 18 Report on the state of cybersecurity in the EU, and NIS2 Article 20 Governance,
- Failure to compliance with approved codes of conduct or approved certification mechanisms,
- Failure in cooperation with the responsible legal person(s) and the competent authorities,
- Failure in measures taken to prevent or limit the damage and/or loss,
- Failure in assessment factor measures like actual or potential financial or economic loss, impact on other services and number of users affected or potentially affected,
- And others.

Accordingly, NIS2 sets out several requirements for high critically and other critically sectors that must be met:

- Regular cyber—risk analyses,
- Information systems security policy (Information Security Management System—ISMS),
- Incident response by handling cyber threat risk incidents and incident management system,
- Business Continuity Management (BCM), e.g., back-up and emergency management,
- Supply chain cybersecurity,
- Cybersecurity training,
- Policies and procedures for the use of cryptography and encryption where applicable,
- Personnel cybersecurity,
- Access control policies and asset management,
- Network and Information Systems resilience,
- And others.

4.3 Liability

Cybersecurity liability is the responsibility of managing directors for cybersecurity mishaps like data breaches and other cybersecurity issues that occur, see NIS2 Article 23 Reporting obligations outlines the obligations for the sectors in Annex I and Annex II to report significant impacting cyber threat risk incidents promptly, ensuring affected parties are informed without increasing the reporting entity's liability. However, it's not a question of if an organization will suffer a breach, but rather when. Therefore, cybersecurity liability assumes responsibility for data

4.3 Liability

breaches and other cybersecurity issues and proactively addresses them by committing to improving organizations cybersecurity posture. This can be achieved through Privileged Access Management solutions (PAM) that support in the detection, response, and recovery from cyber threat risk incidents by controlling and monitoring privileged access, which is critical during incident investigation and mitigation efforts, or encryption, cybersecurity risk assessments, third-party cybersecurity assessments, improved backup plans, and others. Not enforcing these cybersecurity policies result in legal consequences in reporting obligations (NIS2 Article 23 Reporting obligations), legal regulatory fines (NIS2 Article 34 General conditions for imposing administrative fines on essential and important entities, and a loss of customer trust), as illustrated in Tables 4.1 and 4.2.

Table 4.1 Reporting obligations and chains of action

Reporting obligations	Chains of action
Immediately, in any case within 24 h	Upon knowledge of a significant cyber threat risk incident, early warning, indicating, where appropriate, whether there is suspicion that the significant cyber threat risk incident is the result of unlawful or malicious acts, or could have cross-border implications
Immediately, in any case within 22 h	After becoming aware of a significant cyber threat risk incident, report the cyber threat risk incident, updating information and providing an assessment of the significant cyber threat risk incident, including severity and impact, and, if applicable, Indicators of Compromise (IoC)
At the latest within 1 month	Detailed description of the cyber threat risk incident, including severity and impact: – Information on the nature of the cyber threat risk incident or underlying cause that likely triggered the cyber threat risk incident – Information on the remedial actions taken and ongoing – If applicable, the cross-border impact of the cyber threat risk incident

Table 4.2 Legal regulatory fines

Activity	High criticality sectors	Other critical sectors
Supervision through Authorities	Proactive supervision (e.g. regular cybersecurity audits)	Reactive supervision following indications of violations (e.g. targeted cybersecurity checks)
Fines for Violations	Maximum amount of at least €10 million or 2% of global turnover	Maximum amount of at least EUR 7 million or 1.4% of worldwide turnover
Who is included	Large companies from Annex I > €249 million turnover and > €43 million balance sheet Special cases independent of size: e.g. Domain Name System (DNS) service providers, central government, critical infrastructure, and facilities classified as "essential" by the state	Large companies from Annex I > €249 million turnover and > €43 million balance sheet Medium-sized companies from Annex I or Annex II at least 50 employees, or >€10 million turnover & >€10 million balance sheet not a large company Special cases regardless of size: institutions that are classified as "important" by the state

4.4 NIS2 Article 21.2

EU Member States shall ensure that essential and important entities, as listed in NIS2 Annex I and Annex II, take appropriate and proportionate technical, operational and organizational measures to manage the cyber risks, posed to the cybersecurity of network and information systems, which those entities use for their operations or for the provision of their services, and to prevent or minimize the impact of cyber threat risk incidents on recipients of their services and on other services. Thus, NIS2 Article 21 outlines key cybersecurity practices that every organization under the NIS2 Annex I and Annex II regulation must have in place.

4.4.1 Mandatory Cybersecurity Measures

Mandatory cybersecurity measures of NIS2 Article 21.2 are shown in Table 4.3:

4.4.2 Standards in Cybersecurity Risk-Management

With regard to the required cybersecurity measures for NIS2 Article 21.2, shown in Table 4.3, different international cybersecurity standards to cybersecurity risk-management, referred to in Table 4.4.

Table 4.3 NIS2 Article 21.2 security measures

Article 21.2	NIS2 security measures
Policies	Document describing rules, expectations, and overall approach that an organization uses to maintain confidentiality, integrity, and availability of its data, and frequently used in conjunction with other types of documentation like standard operating procedures that work together to enable the organization achieve its security goals. However, an effective security policy must refer to seven elements: • Clear purpose and objectives • Scope and applicability • Commitment from management • Realistic and enforceable policies • Clear definitions of important terms • Tailored to organizations potential risk • Update of purpose and objectives based on lessons learned in relation to changings in the risk landscape Thus, policy concepts related to risk analysis and security for information systems, answering the question *what* and *why*, while procedures, standards and guidelines answer the *how*

(continued)

Table 4.3 (continued)

Article 21.2	NIS2 security measures
Incident Handling	Managing security incidents by recommended actions and procedures needed to do the following: • Recognize and respond to an incident • Assess the incident quickly and effectively • Notify appropriate security team and organizations C-Level of incident • Activate an organization's response • Escalate organizations response efforts based on the severity of incident • Support business recovery efforts made in the aftermath of incident
Business Continuity	Business Continuity Management (BCM) integrates Emergency Management and Response (EMR), Crisis Management and Communication (CMC), Disaster Recovery (DR) (technology continuity) and Business Continuity (BC) (organizational/operational relocation), including testing and updating
Supply Chain	Supply chain security focus on security-related aspects of organizations relationships between different entities, suppliers or service providers to enable low security risks. Reducing exposure to cybersecurity threats and cyber-attack events, management can guard against lost products, downtime, compromised data, lost proprietary information, and loss of customer trust
Security in Acquisition, Development, Maintenance of NIS	Promote security measures and boost EU member states' level of protection and maintenance of critical and crucial network and information systems to improve network and information security of operation in NIS2 Annex I and Annex II sectors, including vulnerability scanning, including newly identified vulnerabilities
Policies & Procedures for Assessing the Effectiveness of Cybersecurity Risk Management Measures	Sets important standards of behavior for cybersecurity risk-management activities because cyberattacks and data breaches are potentially costly. Sets rules how access and assess online applications and internet resources, send data over networks, and information systems, and otherwise to practice responsible security measures
Basic Cyber Hygiene Practices & Training	Including using strong and unique passwords, password management and admin account management and restrictions, regularly updating software and applications, enabling antivirus and firewall protection, avoiding unsafe online behaviors zero-trust principles and identity and access management (IAM). Cybersecurity training focus is on increasing employee knowledge and awareness on potential cyber risks, vulnerabilities and the most common threats that can negatively impact on organizations business

(continued)

Table 4.3 (continued)

Article 21.2	NIS2 security measures
Policies & Procedures for Assessing Effectiveness of Cybersecurity Risk Management Measures	Plan of action enabling to prevent and remediate security risk events that could occur on organizations information system, based on identifying, analyzing, evaluating, prioritizing, managing and monitoring risks to organizations information systems to effectively managing cybersecurity risks. Often this lack of two-factor authentication or the holistic perspective necessary to address cyber risk in a comprehensive and consistent manner, which is required due to today's risk landscape
Personnel Security, Access Policies & Asset Management	Aims to manage and minimize the risk of staff exploiting legitimate access to an organization's assets or premises for unauthorized purposes. These purposes can encompass many forms of cyber-criminal activity, from minor theft through to advanced cyber-attacks
Policies & Procedures Regarding the use of Cryptography & where applicable, Encryption	Encryption keys covered by this policy must be protected to prevent their unauthorized disclosure and subsequent fraudulent use. Keys used for secret key encryption, also called symmetric cryptography, must be protected as they are distributed to all parties that will use them. Encryption policies define when encryption should or shouldn't be used and encryption technologies or algorithms that are acceptable
Use of Multi-Factor Authentication (MFA) or Continuous Authentication Solutions, Secure Communication for Voice, Video, & Text, & Secure Emergency Communication Systems within the Entity	Multi-factor authentication (MFA) is a multi-layered security system that verifies the identity of users for login or other transactions. By leveraging multiple authentication layers, even if one element is damaged or disabled, the user account will remain secure. Continuous MFA through passive biometrics is the only solution capable of checking users' identities hundreds of times each day without requiring any participation from the users—making it secure, efficient, and compliant End-to-End-Encryption and data-centric security concepts, access policy and access management, and automated access decisions (NIS2 Paragraph 98). IAM secures and controls access based on policies Emergency communication systems are networks that allow individuals to exchange information during crises. They can also aid communication if normal channels temporarily fail

Tables 4.3 and 4.4 illustrates that managing cyber threat risk incidents of high critical and other critical entities apply specific procedures, strategies and standards that enable to cover a variety of essential cybersecurity measures. Hence, for cybersecurity risk-management specific measures arise that have to be applied. These specific measures are, for example, in accordance to Article 21.2.a:

- Identify potential cybersecurity risks as part of a vulnerability analysis,
- Assess potential identified cybersecurity risks in terms of their impact on cybersecurity,

4.4 NIS2 Article 21.2

- Use efficient and effective programs/tools for assessment,
- Use maturity/capability models as efficient measure for increasing cybersecurity efficiency and quality in organizations,
- And others.

Table 4.4 NIS2 Article 21.2 international standards used in risk-management

Article 21.2	International security standards in risk-management
(a)	**NIST CSF 2.0:** Set of cybersecurity best practices and recommendations, providing a voluntary, actionable guidance enabling organizations to manage, reduce, and communicate cybersecurity risks for networks, information systems, and others **NIST SP 800-53:** Set of recommended cybersecurity and privacy controls for federal information systems and organizations support meeting Federal Information Security Management Act (FISMA) requirements **ISO/IEC 27000:** Family of standards, also known as Information System Management System (ISMS) family of standards or simply ISO27K, which cover a broad range of information security standards published by both the International Organization for Standardization and International Electrotechnical Commission **ENISA Risk-Management/Risk Analysis:** Input to decision makers whether cybersecurity risks need to be treated or not and what is the most appropriate and cost-effective cybersecurity risk treatment methodology
(b)	**NIST SP 800-61:** Guidelines for organizations on how to handle cybersecurity events **ISO/IEC 27036-2:2023:** Multi-part standard providing guidance on managing information security risks associated with acquiring goods and services from suppliers, particularly ICT products
(c)	**NIST SP 800-34:** Contingency Planning Guide for Federal Information Systems, to provide instructions and recommendations for information technology systems contingency planning **NIST SP 800-184:** Enables federal agencies in a technology-neutral way in improving their cyber threat risk incident event recovery plans, processes, and procedures **NIST SP 800-209:** Provides an overview of the development and evolution of storage technology, examines current data storage cyber threat risk incidents, and provides a detailed set of cybersecurity recommendations and guidance to address storage cybersecurity **ISO/IEC 22301:2019:** Specifies the structure and requirements implementing and maintaining a Business Continuity Management System (BCMS) that develops BC appropriate to the amount and type of impact that the organization may or may not accept following a disruption **ISO/IEC 27031:2011:** Describes concepts and principles of Information and communication technology (ICT) readiness for business continuity (BC), and provides a framework of methods and processes to identify and specify all aspects (performance criteria, design, and implementation) for improving an organization's ICT readiness to ensure BC **ISO/IEC 27040:2015:** Provides detailed technical guidance on how organizations can define an appropriate level of cyber risk mitigation by employing a well-proven and consistent approach to planning, design, documentation, and implementation of data storage cybersecurity

(continued)

Table 4.4 (continued)

Article 21.2	International security standards in risk-management
(d)	**NIST SP 800-161**: Guidance on identifying, assessing, and responding to cybersecurity risks throughout the supply chain at all levels of an organization **ENISA Interoperable EU Risk-Management System:** Describes and evaluates interoperability features for risk-management frameworks and methods, by employing a four-level scale to evaluate their interoperability level
(e)	**NIST SP 800-100:** Information Security Handbook providing a broad overview of information security program elements to assist managers in understanding how to establish and implement an information security program. **NIST SP 800-123:** Guide to General Server Security contains recommendations on how to cybersecure servers, offering general advice and guideline on how to approach this mission **NIST SP 800-137:** Information Security Continuous Monitoring (ISCM) for Federal Information Systems and Organizations, provides guidance and best practices for establishing, implementing, and maintaining a continuous monitoring program for information security in federal agencies and organizations **NIST SP 800-216:** Recommendations for Federal Vulnerability Disclosure Guidelines, a flexible, unified framework for establishing policies and implementing procedures for reporting, assessing, and managing vulnerability disclosures for systems within the Federal Government **ISO/IEC 29147:2018:** Provides requirements and recommendations to vendors on the disclosure of vulnerabilities in products and services. Vulnerability disclosure enables users to perform technical vulnerability management as specified in ISO/IEC 27002:2013 **ENISA Coordinated Disclosure of Security Vulnerabilities:** Ensure that vulnerabilities are disclosed to the public after the responsible parties developed a fix, a patch or provide mitigation measures to limit the cyber threat risk incident posed by the exploitation of a vulnerability **EU Cyber Resilience Act (CRA):** Lead to the development and implementation of common cybersecurity standards in the EU for products with digital elements
(f)	**NIST SP 800-55:** Advises organizations to design their performance measurement programs to support the selection and implementation of cybersecurity controls **MITRE ATT&CK:** Globally-accessible knowledge base of adversary tactics and techniques based on real-world observations. ATT&CK knowledge base is used as foundation for the development of specific cyber risk incident models and methodologies in the private sector, in government, and in the cybersecurity product and service community **ENISA National Cybersecurity Assessment Framework (NCAF):** Measure maturity level of cybersecurity capabilities of EU Member States to support them in conducting an evaluation of their national cybersecurity capability, enhancing awareness of the country maturity level, identifying areas for improvement and building cybersecurity capabilities

(continued)

4.4 NIS2 Article 21.2

Table 4.4 (continued)

Article 21.2	International security standards in risk-management
(g)	**NIST SP 800-16:** Approach to information security awareness training and role-based training at lower tactical level than 800-50 **NIST SP 800-50**: Document provides guidelines for building and maintaining a comprehensive awareness and training program, as part of an organization's IT security program. The guidance is presented in a life-cycle approach, ranging from designing, developing, and implementing an awareness and training program, through post-implementation evaluation of the program. It includes guidance on how IT security professionals can identify awareness and training needs, develop a training plan, and get organizational buy-in for the funding of awareness and training program. Approach is at higher strategic level than 800–16, provides guidelines for building and maintaining a comprehensive awareness and training program, as part of an organization's IT security measure **NIST SP 800-53:** See under (a) **NIST 1800:** Specific cybersecurity challenges in the public and private sectors. Practical and user-friendly guide that facilitate adoption of standards-based approaches to cybersecurity **NIST NICE Framework:** Establishes a common language that describes cybersecurity work and knowledge and skills needed to complete that work **NISTIR 7621:** Interagency report (IR) that present the fundamentals of a small business information security program in a non-technical language **ENISA ECSF:** Summarizes cybersecurity-related roles into 12 profiles, which are individually analyzed into the details of their corresponding responsibilities, skills, synergies and interdependencies
(h)	**NIST CSF:** See under (a) **NIST SP 800-175A:** Guideline for Using Cryptographic Standards in the Federal Government: Directives, Mandates and Policies provides guidance on the determination of requirements for using cryptography **NIST SP 800-175B:** Provides guidance to the Federal Government for using cryptography and NIST's cryptographic standards to protect sensitive but unclassified digitized information during transmission and while in storage. The cryptographic methods and services to be used are discussed **ISO27001 Family of Standards:** See under (a) **ISO/IEC 27001:** Promotes a holistic approach to information security: vetting people, policies and technology. An ISMS implemented according to this standard is a tool for **risk-management, cyber-resilience** and **operational excellence** **ENISA Securing Personal Data:** Focus on how to engineer data protection in practice and puts forward analyses, recommendations and relevant use cases on how (cybersecurity) technologies and techniques can support the protection of engineering data and the fulfilment of the General Data Protection Regulation (GDPR) data protection principles. GDPR distinguishes between personal data and sensitive personal data. Personal data are any information relating to an identified or identifiable natural person (data subject), while sensitive personal data is a specific set of special categories

(continued)

Table 4.4 (continued)

Article 21.2	International security standards in risk-management
(i)	**NIST SP 800-50:** Provides guidelines for building and maintaining a comprehensive awareness and training program, as part of an organization's IT security program **NIST SP 800-192:** Document and reviews methods for the verification for access control models and the testing of model implementations, and define structures for AC models, and demonstrate expressions of AC models and safety requirements in a specification language of a model checker. The document showed the use of black box and white box model checkers that verify the integrity, coverage, and confinement of the specified safety requirements against models **NIST SP-1800-6:** Describes a cybersecurity platform for trustworthy email exchanges across organizational boundaries, including reliable authentication of mail servers, digital signatures, encryption of email, and binding cryptographic key certificates to sources and servers **NISTIR 7316:** In addition to authentication mechanism, access control is concerned with how authorizations are structured. In some cases, authorization mirror the structure of the organization, while in others it be based on sensitivity level of various documents and the cybersecurity level of the user accessing those documents. Explains some of the most commonly used access control services available in information technology systems, their structure, where they are likely to be used, and advantages and disadvantages of each. **NISTIR 7874:** Guidelines for Access Control System Evaluation Metrics, with background information on Access Control (AC) properties, and enable AC experts improve their evaluation of highest security AC systems by discussing the administration, enforcement, performance, and support properties of AC mechanisms that are embedded in each AC system **ISO/IEC 19770:** Framework of IT asset management (ITAM) processes to enable an organization to prove that it is performing software asset management that meets corporate governance standards **ISO/IEC 29146:2016:** Defines and establishes a framework for access management (AM) and cybersecure management of the process to access information and Information and Communications Technologies (ICT) resources, associated with the accountability of a subject within some context **ENISA's AR-in-a-Box:** Comprehensive solution for cybersecurity awareness activities designed to meet needs of public bodies, operators of essential services, and both large and small private organizations. It provides theoretical and practical knowledge on how to design and implement effective cybersecurity awareness programs

(continued)

4.4 NIS2 Article 21.2

Table 4.4 (continued)

Article 21.2	International security standards in risk-management
(j)	**NIST SP 800-34:** Refers to Contingency Planning interim measures to recover IT services following an emergency or system disruption. **NIST SP 800-45:** Recommend cybersecurity practices for designing, implementing, and operating email systems on public and private networks, containing information on popular email encryption standards and other standards relating to email. It presents general information on securing mail servers' operating systems and specific guidance on cybersecure mail server applications, protecting messages traversing servers, and cybersecure access to mailboxes. It also provides information regarding email client cybersecurity and mail server administration **NIST SP 800-63B:** Guideline focus on authentication of subjects interacting with information systems over networks, establishing that a given claimant is a subscriber who has been previously authenticated. **NIST SP 800-84:** Assist IT personnel responsible for designing, developing, conducting, and/or evaluating Test, Training and Exercise (TT&E) events in fulfilling these responsibilities effectively, enabling organizations to identify problems related to an IT plan and implement solutions before a cyber threat risk incident situation occurs **NIST SP 800-170:** Refers to authentication using two or more different factors to achieve authentication. Factors include PIN, password, cryptographic identification device, token, biometric, and others **NIST SP 800-177:** Address importance of content cybersecurity in communications, providing guidance on implementing measures such as email content filtering, antimalware scanning, and Data Loss Prevention (DLP) techniques **ISO TR 29156:2015**: Provides guidance specifying performance requirements for authentication using biometric recognition in order to achieve desired levels of cybersecurity and usability for the authentication mechanism

In this context, the Cybersecurity Maturity Level Model (CMLM) determine the maturity level of the overall organization's cybersecurity maturity (see Sect. 3.3). The main focus is not on whether individual processes are being carried out or are incomplete. The starting point of a CMLM is referred to as initial. For more details see Chap. 3.

With regard to Cybersecurity Capability Maturity Integrated (CMMI) models, they are the representation in capability levels to reflect a specific process area or behavior to be improved and assign the desired capability level to it. Thus, it is important to note whether a process is being carried out or is incomplete, has an impact on the starting point, which could be, for example, referred to as incomplete. For more details see Chap. 3.

An essential focus increasing the maturity or capability level is represented by the cybersecurity awareness level of employees, as key component of information security solution. Given that fact, a cybersecurity awareness and training program for employees is vital to a cybersecurity solution to achieve expected cybersecurity goals. The importance of cybersecurity awareness and training of employees as a countermeasure against cyber risks cannot be overstated, and need the development of cybersecurity training in an organization wide training program. As introduced in [19], the following must be considered when building a cybersecurity and privacy learning program:

- Implement a life cycle model that allows for ongoing, iterative improvements and changes to accommodate cybersecurity, privacy, and organization-specific events,
- Implement a learning program concept that incorporates features found in IST documents,
- Leverage current NIST guidance and terminology in reference documents, such as the NICE Workforce Framework for Cybersecurity, the NIST Cybersecurity Framework 2.0, the NIST Privacy Framework, and the NIST Risk-Management Framework,
- Implement an employee-focused cybersecurity and privacy culture to organizations,
- Integrate learning programs with organizational goals to manage cybersecurity and privacy risks,
- Address the challenge measuring the impacts of cybersecurity and privacy learning programs.

4.5 Preparing for NIS2

NIS2 Directive aims to establish a higher level of cybersecurity and resilience within the organizations of the EU. Therefore, NIS2 brings into scope more sectors and focuses on providing guidelines to ensure uniform transposition in local law across EU Member States. Thus, organizations must prepare for NIS2 by defining their compliance roadmap and optimizing their cybersecurity awareness. Necessary actions and measure to be carried out shown in Table 4.5.

Risk mitigation activities and controls required to fulfill NIS2 requirements are:

- **Incident Management:** Prevention, detection and response to cybersecurity incidents to restore the regular operational state.
- **Business Continuity and Crisis Management w.r.t. Recovery and Response to Cyber Threat Risk Incidents:** Business Continuity Plan (BCP) ensures that organization can resume operations as swiftly as possible, minimizing downtime and associated losses (financial and reputational), and crisis management focus on handling unexpected cyber threat risk incidents.

4.5 Preparing for NIS2

Table 4.5 Preparing for NIS2

Action	Measures
Relevance/Concern	Clarify whether organization belongs to the sectors listed in NIS2 Annex I and Annex II Organizations will not informed whether they are affected by NIS2 Organizations should independently assess own NIS2 impact based on specified sectoral and size-related aspects Clarify whether organizations customers belong to essential and important sectors in NIS2 Annex I and Annex II
Responsible	Clarify who from management is operationally responsible for implementing NIS2 measures Responsible team in organization implement, document security measures to prevent and mitigate damage in order to reduce any possible impacts (if necessary, bring in external consultants), and finally report to management
Reporting Obligations	Clearly define responsibility: Changes to registration data (Article 3 Essential and important entities): If registration data stored with national competent authorities changes, changes must be submitted within 2 weeks Significant cybersecurity incident (Article 23 Reporting obligations): If a cybersecurity incident falls into this category, it must be reported to national authorities and, if applicable, customers or users of the organizations own services Voluntary reports (Article 30 Voluntary notification of relevant information): In addition to reporting obligations, there is the option for companies to voluntarily report cyber threat risk incidents, regardless of whether they are affected by NIS2: – Cybersecurity incidents that are not significant, cyber threats risk incidents and near-misses – Organizations not affected by the directive in the event of significant cybersecurity incidents, cyber threat risk incidents and near-misses – Vulnerabilities can be reported by individuals/organizations—even anonymously—to the relevant national authorities
Identification	Determine organizations assets based on specific risk analysis of affected facilities/areas/processes and delta to NIS2
Risk Analysis	Cover core elements that strengthen the prevention and management of cybersecurity incidents, including implementation of risk analysis and cybersecurity concepts, incident management and ensuring cybersecurity in supply chains, maintaining business operations, access control, Multi-Factor Analysis (MFA), continuous authentication, emergency communication solutions, and more. Evaluation of assessment (GAP analysis): Delta status company and NIS2 to create implementation plan, align GAP analysis with ISO/IEC 27001 Determination of status quo maturity level/capability level in cybersecurity based on CMLM/CMMI (internal or external assessment) and measures to ensure maturity level/capability in relation to NIS2 recommendations for cybersecurity measures
BC/BCM	Ensure Business Continuity (BC) even in the event of major cybersecurity incidents. A well-developed and practiced Business Continuity Management (BCM) is a prerequisite for this (RPO, RTO, MtD). This includes established emergency procedures and resilient organizational structures. Appoint the person responsible for reporting obligations

- **Vulnerability Management (Detection, Prevention and Remediation):** Includes pen-testing if necessary (an authorized simulated attack performed on a digital system to evaluate its cybersecurity), third-party risk (likelihood that organization experience an adverse event such as data breach, operational disruption, reputational damage, when organization choose to outsource certain services or use software built by third parties to accomplish certain tasks).
- **Supply-chain Security Management (Internal and External:** Internal supply chain cybersecurity management refers to organization's cybersecurity rules with customers, and external supply chain cybersecurity management that includes cybersecurity rules with a network of suppliers, manufactures and other partners, often complex with many different stakeholders involved.
- **Network and Information System Security:** Is a subset of cybersecurity, focused on protecting an organization's IT infrastructure from online cyber threat risk incidents, including tools and processes for preventing, detecting, and remediating cyber threat risk incidents to sensitive information, also concerned with documenting processes of cyber threat risk incidents that affect the cybersecurity.
- **Cybersecurity in Development (Access Controls, Encryption, …):** Tools, practices, and approaches to identify and prevent cybersecurity flaws during early development of information/software systems, when it is most cost effective by incorporating cybersecurity into every phase of the System Development Life Cycle (SDLC),
- **Cybersecurity Training:** For employees to capture cybersecurity awareness and improve information management and compliance behaviors and mitigate the risk of cyber threat risk incidents.

4.6 Business Continuity Plan

Business Continuity (BC) refers to processes and strategies organizations should put in place to ensure that critical business assets and operations can continue during and after a disruption event. Hence, organizations should have a plan to deal with difficult situations, so the organization can continue to operate despite disruptions. Whether it's a business, public sector organization, or other, it is important to know how to keep going under any circumstances. In this context, the business continuity plan (BCP) recognizes potential cyber threat risk incidents to an organization and analyses what impact they may have on day-to-day business operations. It also provides a way to mitigate these cyber threat risk incidents, putting in place a framework that allows key functions of the business to continue even if a cyber threat risk incident happens. Thus, a BCP is a comprehensive plan and encompasses all aspects of an organization, including IT systems, personnel, facilities, and supply chains with the primary focus on identifying essential business functions, and the resources needed to restore them swiftly.

Creating a BCP can be either straightforward or more involved, depending on the size and complexity of the business and the cyber risk the organization face, and involves activities such as [15]:

- Business cyber risk level and impact analysis,
- Documenting activities necessary to prepare the organization for possible cyber risks, including strategic recovery measures,
- Identifying and authorizing detailed activities for any cyber risk disaster recovery phase,
- Identifying and authorizing detailed activities and measures for managing the business recovery process,
- Testing and auditing the business recovery process to act during and after the possible cyber risk crisis,
- Training business recovery process activities and measures, the required support technologies and communications by the CEO or COO to keep employees and staff aware and well-informed,
- Implementing a process for keeping the BCP up to date.

Either way, a BCP is a detailed strategy with a set of actions to enable an organizations ability to prevent the rapidly recover from a significant disruption to its business operations. Core components of a set of actions contain:

- Cyber Risk and impact analysis,
- Recovery strategy with Recovery Time Objective (RTO) and Recovery Point Objective (RPO) measures in order to minimize downtime and data loss,
- Team assignments with roles and responsibilities in BCP, clearly define their tasks, decision-making authority, and communication channels.
- Communications Guidelines,
- Regular Testing and Training.

Developing a successful BCP evolve several questions to be answered such as:

- What is the BCP objective, or why does the organization need it?
- What constitutes a disaster that would activate the BCP?
- Who does what during a potential occurring disaster?
- How will personnel communicate, and who contacts whom?
- What is the likelihood of various cyber threat risk incidents and human errors?
- What is the organizations business impact of each of those events?
- What technologies is the organization leveraging to ensure continuity?
- What weaknesses and gaps do the organization need to correct and fill?
- And others.

Answering these questions in a conscious way, the BCP issues a critical and crucial background that executives, stakeholders, and personnel know what to do and how to do it, and can easily access the BCP and follow the outlined steps, if any confusion happens.

The next activity mandates the recovery team to immediately manage the BCP. The recovery team ideally consist of IT personal and employees from other business-critical units of the organization. They play key roles in organizing and carrying out the emergency procedures and should have knowledge and authority to make decisions without support from supervisors. Their responsibilities include:

- Documenting and updating the BCP NIS2 conform,
- Identifying potential cyber threat risk incidents and establish preventive solutions for backups after a potential threat incident
- Training personnel on disaster response actions,
- Coordination of organization-wide communication requires specifying who's in charge of updates, which channels organization should use, how often organization should communicate, and others. It also should include how to handle communication with media and authorities to keep organizations message consistent and accurate,
- Activating the BCP when a situation warrants it.

4.6.1 BCP Component Risk and Impact Analysis

The BCP core component, risk and impact analysis, involve identifying potential cyber threat risk incidents that might disrupt organizations business, and estimating how likely they be, as well as consequences to be happen. To this purpose, a risk assessment must be carried out to identify possible cyber threat risk incidents. Based on this a recovery timeline and prioritizing potential efforts must be planned to streamline BCP. During risk assessment the potential cyber threat risk incident that pose the biggest risk to the organization be identified. Risk assessment covers features such as:

- Nature of organizations business and decision whether organization handle crucial and/or valuable data,
- Decision about structural or site-specific vulnerabilities, or cybersecurity threat and/or malicious cyber-attack events,
- Likelihood of human-caused events, including internal human cybersecurity errors and/or not available cybersecurity awareness,
- Assessing potential impacts of business downtime or data loss on organizations processes,
- Potential risk through third party business units,
- Technology inventory to conduct an audit of IT assets to obtain a current overview of hardware, software, cloud services, external service providers and other resources, essential for business operations, to enable effective and efficient risk-management and appropriate disaster recovery planning,
- And others.

Another component of risk assessment is the Business Impact Analysis (BIA). This explores how the identified potential cyber threat risk incident will affect the organizations business. On this basis one can plan strategically and prioritize needed

4.6 Business Continuity Plan

resources appropriately. For most businesses, the impact of a malicious cyber threat risk incident, e.g., a ransomware attack, that has both high likelihood and impact, e.g., a financial and a reputational disaster, is based on several factors:

- Business operations that will affect and in what way,
- How long the outage will last,
- Number of employees the event will idle, and for approximately how long,
- Whether it will affect revenue, ransom and organizations trustability,
- Estimated costs of recovery,
- And others.

Using each of these points, organization can calculate true cost of a malicious cyber threat risk incident in monetary terms of hourly or daily losses as well as probable loss in trustability to business partners.

4.6.2 BCP Component Backup Schedule

The BCP core component, recovery strategy, define and document the scope of the BCP for critical business functions, data and resources with documentation of personal roles and responsibilities for the disaster recovery strategy that enables organizations to continue to work successfully despite a cyber threat risk incident situation. For this purpose, backup schedules are set up in line with Recovery Point Objective (RPO)and Recovery Time Objective (RTO), in order to minimize downtime and data loss and swiftly restore data and systems after a disruption happen. In addition, the role and responsibilities according to team assignments are activated to run the disaster recovery strategy according to plan, in order to keep the organizations operation despite the cyber threat risk incident situation impact continuous. A list of disaster response procedures might include the following actions:

- Notify recovery team leads and senior management of the scope of the disaster event situation,
- Diagnose affected devices, processes, servers, applications, and others
- Contact appropriate vendors during an application outage or disaster event situation affecting third-party systems or recovery tools,
- Establish measures and actions to enable timely backup that minimize downtime and data loss based on RPO and RTO,
- Notify insurance provider(s) if applicable,
- And others.

Based on the BCP disaster response procedures, RPO, RTO and MtD (Maximum tolerable Down-time) are thresholds for evaluating data security, disaster recovery and business continuity activities. With RPO, RTO, MtD, a high availability configuration is established that ensures that IT systems remain operational within specified limits in the event of routine problems or possible failures, which is part of a Disaster Recovery Planning (DRP). A DRP is a policy designed to enable an organization to comprehensively execute consistent recovery processes in response to a disaster event to protect business continuity. Thus, a DRP takes over with replacement of devices and restoration of data from back-ups to enable critical business functions within the thresholds (restore) of RPO, RTO, and MtD.

Recovery Point Objective (RPO)

RPO is defined as the maximum amount of data, measured by time, which can be lost after a recovery from a disaster event, failure, or comparable event before data loss will exceed what is acceptable to an organization. RPO also determines the maximum age of the data or files that must be recovered from backup storage devices in order to resume regular business operations after a successful cyber threat risk incident, to meet the objectives specified by the RPO. Furthermore, RPO is expressed in a time span, i.e., the point in time before a cyber treat risk incident, when data/file can be successfully recovered. This is the time elapsed since the last reliable backup, which can be specified in seconds, minutes, hours or days. RPO can be derived based on the following questions:

- How often is data updated?
- What effort would be required for the user to reconstruct, if possible, the data created or updated since the last backup?
- If recovery is not possible, how much current data can a cyber-attacked organization permanently lose, considering the probability of catastrophic data loss?
- And others,

to determine

- How much data/files will be lost after a cyber disaster event,
- How frequently backup data for disaster recovery purposes is needed, in other words, RPO does not concern other IT needs [20].

Calculating RPO for a business organization considers several steps with specific objectives, as shown in Table 4.6 [16, 17].

From Table 4.6 one can see that the frequency data or files are updated (stored) depends on the RPO needs, to ensure restore operations with minimal data loss following a disruptive event. Factors that can affect RPOs include [20]:

- Maximum tolerable data loss of the specific organization,
- Industry-specific factors such as businesses dealing with sensitive information such as financial transactions or health records must update more often,
- Data storage options, such as physical files versus cloud storage, can affect speed of recovery,
- Cost of data loss and lost operations,

Table 4.6 Calculating RPO [16, 17]

Step	Calculation activity	Objectives
1	How often files updated	Set up RPO to match update frequency to ensure most up-to-date information is retrievable For example: Digital files and transactions update every 30 min to ensure continuously access to recent information with minimal data loss
2	Review BCP goals	RPO support BCP goals, review elements care-fully to determine separate RPOs of various time allotments for business units For example: Financial transactions and critical data processes of banks are vital, require shorter RPO times than other resource files of personnel records, which get updated less frequently and can sustain a longer RPO time
3	Consider Industry Standards	Industry with unique RPO needs can consider industry standards calculating RPOs for business units. Common intervals for RPO are: **0–1 h:** Shortest time frame for critical operational elements that can't afford to lose an hour of data, typically because high volume, dynamic or difficult to recreate For example: Online banking transactions, patient records or stock market trading activities **1–4 h:** For business units deemed semi-critical, only some data loss is acceptable For example: Customer online chat logs, file servers or social media records **4–12 h:** For business units that update data daily or less frequently, like advertising and marketing, sales or operational statistics data. Typically, these units rarely have a grave impact on a business if affected For Example: An outage occurs. If the RPO is 12 h and the last copy of data available is from 10 h ago, it is still within the RPO's parameters for this BCP **13–24 h:** Setting longer RPO timeframe for important, but not critical data and business units rarely exceed 24 h. These settings used for actions like purchase orders, inventory control or personal files
4	Establish and Approve RPO Settings	After factoring all concerns to each element of data management, establish RPOs and have approve them by leadership for implementation by IT teams or business partners to implement. Properly document the process and keep the records to refer to or use as baseline when reassessing or adjusting. Depending on the role in organization, use existing RPO processes and procedures or create or establish them
5	Analyze RPO Settings Consistently	As organization grows or BCP change, so might the RPO objectives. Consider establishing a routine and frequent review and analysis how good existing RPO settings perform and if any need adjusting. Important step even though it comes last. If a data loss or system failure happens, an ad hoc review and in-depth analysis of how RPO and RTO performed can help to learn meaningful insights into the overall data management system

- Compliance schemes include provisions for disaster event recovery, data loss, and data availability that may affect organizations businesses,
- Cost of implementing disaster event recovery solutions.

Once defined, RPOs serve to detail the goals of the BCP, and organizations business unit should have distinct RPOs. For example, financial transactions and other mission critical data processes demand shorter RPOs than less frequently updated files such as personnel records [20].

Recovery Time Objective (RTO)

RTO is the amount of time an organizations business has to restore its processes at an acceptable service level after a disaster event to avoid intolerable consequences associated with the disruption. Thus, RTO (Recovery Time Objective) refers to the maximum acceptable duration for which the attacked devices, systems or applications fail after a successful cyber threat risk incident. It also defines the time span between the occurrence of the cyber threat risk incident and the restoration of regular business operations, as well as the amount of downtime costs incurred per unit of time due to the cyber threat risk incident. In this regard, for each device, system, soft-ware application, database and others, the IT department must examine the Maximum tolerable Downtime (MtD) applicable to it, and create an assessment of the dependencies of the IT components, and work through these dependencies backwards to determine how much time will be spent restoring each technical component. In addition, redundancy, repair times, and the availability and competences of the assigned staff are important. In this context, the question is how to calculate RTO. RTO can be calculated by figuring the outer limits of an outage's length that can be tolerated by the organizations business. To get to this in-formation, answering the following questions is essential:

- How much time after notification about the organizations business process disruption can it take to resume normal operations?
- What service level agreement (SLA) has the organization in place for users of the systems (internal, external, multilevel)? SLA refers to a document that outlines a commitment between a service provider and a client, including details of the service, the standards the provider must adhere to, and the metrics to measure the performance. Typically, IT departments use service-level agreements.
- Is the system attacking a customer-facing system? How would it affect the customer experience, loyalty, and churn if unavailable?
- How much revenue would the organization lose if the systems were not available?
- Would other systems be impacted if one was offline? How critical are they? What is their SLA and RTO?
- And others.

RTO is often used in conjunction with RPO to enable organizations businesses determine how often they should schedule data backups. If RTO is 12 h for a particular system, then the system must be restored within that period. Failure to restore the system in time could have disastrous consequences for the normal business operations. As per ISO 22301, RTO is defined as the maximum allowable time that a business can tolerate for the recovery of its critical activities after an incident [21], but it can vary depending on the application, system, or organization. During the process of BCP, RTO timelines are determined through Business Impact Analysis (BIA). The BIA purpose is to correlate the system with the organizations critical business processes and services provided, and based on that information, characterize the consequences of a disruption. In addition, this analysis enables assess the potential impact on the organizations business and enables effective decision-making for timely Disaster Recovery (DR) and Business Continuity Planning (BCP) implementation.

The recovery process timeframes can range from seconds to hours or even days. An important objective of RTO is to assess the time required to restore and resume normal business operations after a major cyber threat risk incident. However, it is important to note that the shorter the RTO, the more expensive the disaster recovery plan. By identifying their RTO, organizations business units can define their DR processes, including data backup and restore, data protection, and allocate resources accordingly to ensure Business Continuity (BC) [21]. To calculate RTO, a scheme is given in [21]:

- **Identify RPO:** Identify RPO means maximum length of data loss that organization is willing to lose in a successful cyber threat risk incident or system downtime. Then the organization be able to recover any critical data that was lost at maximum within this length.
- **Calculate Backup Window:** Amount of time it takes for all data backup to be stored securely offsite. This also includes full data backups and incremental backups, as well as any time needed for validation and verification processes.
- **Estimate Data Restoration Time:** Having CBW it can be estimated how long it will take to restore all lost data in a cyber threat risk incident or system failure. This time can vary depending on the type of data being restored, as well as the number of servers and databases that need to be restored.
- **Calculate Data Processing Time:** Calculation on how much time it will take for any necessary processing and analysis tasks if data has been restored. This includes tasks such as validating new records or running reports against newly restored databases.
- **Estimate Testing Time:** Once all data has been restored and processed, it's important to test that everything has been successfully recovered before going live with new systems or applications. Depending on the complexity, this testing process could take anywhere from several hours to several days or even weeks.
- **Determine RTO:** Once all of these times have been estimated, determine what the overall RTO should be for any given cyber threat risk incidents or system failure scenario of organizations business-critical applications. With regard to this fact, RTO should include enough buffer time so that there is no risk event of losing any critical customer or business information due to an unexpected delay in recovery efforts.

The main difference between RTO and RPO is that RTO is the amount of time it takes to recover after a disaster event strikes, while RPO refers to the amount of data loss organizations business can tolerate without causing severe damage. In other words, RTO defines the maximum acceptable downtime for an application or service, while RPO sets the maximum acceptable data loss that may occur.

Maximum Tolerable Downtime (MtD)

MtD is the maximum duration a system or device can remain inoperable before causing severe impact on an organization. Hence, it represents the total time needed to transition from disaster event recovery to normal business operations. Therefore, MtD covers the period of time that applications or data may be unavailable to a particular user, whereby the threshold must be set by the company management (C-level), as it is part of DRP and BCP at the executive level. Per se, technical functionality must be immediately recoverable after a cyber-attack event or a system crash, as users must catch up on work missed during the outage while the MtD period is still running. In this MtD is an important metric of any Business Continuity Plan (BCP). When this metric don't met, a business is likely to suffer negative impacts to operability, revenue, and organizations reputation. To calculate MtD, BCP planning teams often sum three metrics: Incident Response time (IRT), Recovery Time Objective (RTO), and Operational Resumption Time (ORT), as follows:

$$MtD = IRT + RTO + ORT$$

as necessary measures that determine to restore acceptable business operation following a disruption event like a successful cyber threat risk incident [22]. Thus.

- **Incident Response Time (IRT):** Represents the elapsed time from when a disruption event occurs to when a disaster cyber risk event is declared and the DR plan activated. For minor to moderate disruptions, the IRT might be able alone to manage the cyber threat risk incident. For more severe disruptive events, DR teams execute the DRP. The speed at which a disruption event is detected depend on the effectiveness of the monitoring or alert system. The speed of disruption event assessment and DRP activation depends on staff training and the cyber threat risk incident response workflows involved.
- **Recovery Time Objective (RTO):** Measures the time required to resume operating the technology infrastructure at a normal or acceptable level.
- **Operational Resumption Time (ORT):** Describes all other activities needed to restore business functionality after IT systems are up and running that can also include tasks required to get applications running properly. Comprehensive understanding of a business process is required to accurately estimate ORT and whether it can be met during a disruption, which is a key aspect of Business Impact Analysis (BIA).

4.6 Business Continuity Plan

Cost Calculation of Successful Cyber Threat Risk Incidents

Recent research pegs the average cost of a successful cyber threat risk incident at a staggering $4.45 million, which represents the sum of multiple factors, from the cost of an initial response through to the lost associated with lost IP or stolen data. For example, ransomware significantly adds to the cost of a cybersecurity data breach. Some estimates put the average ransom payment in 2023 in the hundreds of thousands of dollars up to over half a million dollars. Despite officials recommend with organizations to disregard hacker demands, 56% of ransomware affected opted to pay up in exchange for getting their data back. Organizations may agree to pay simply because the ransom is less than the combined costs of system downtime, reputational harm, and non-compliance fees that come from a publicly disclosed data breach. But paying off a cybercriminal to avoid extensive repercussions doesn't always work [23].

The following simple example of cost calculation of a ransomware attack in the context of recovery costs after a ransomware attack according to the three-factor model, shown in Table 4.7, illustrate the expected costs.

Based on the result of Table 4.7 it can concluded that if the ransom is less than $3 million, the organization should pay ransom because the own cost is higher and the organization also comes back to own work later on. If the ransom is higher, the organization has to decide the pros and cons to solve the inequality that refers to select a decision

Ransom Cost *is smaller or equal* Recovery Cost
Ransom Cost *is bigger or equal* Recovery Cost

Table 4.7 Recovery cost example of a ransomware attack

Loss of productivity factor	Cost example [$]
Average daily compensation (salary + benefits)	3000
Downtime (days)	15
Number of employees unable to work due to ransomware	50
Subtotal	22,500,000
Loss of Revenue Factor	**Cost Example [$]**
Average daily turnover	75,000
Downtime (days)	15
Subtotal	1,125,000
Costs for Restoring IT Operations Factor	**Cost Example [$]**
Overtime employee or wage contractor restoration of operations	350,000
Hardware and Software Repair	150,000
Subtotal	500,000
Total base cost of ransomware incident	3,875,000

Business Continuity Planning Communication Guidelines

An essential BCP core component is communication guidelines. When cyber-attack events happen, everyone on the assigned team members to be on the same page. This is why having effective communication guidelines is a must for creating a solid BCP. The communication guidelines define how team members share information with assigned teams, employees, stakeholders, business partners, NIS2 cybersecurity law directive responsible, the public and media. Thus, it is important to specify who's in charge of updates, which channels should be used, and how often the organization should communicate. BCP also, include directives how to handle communication with the media and authorities to keep messages consistent and accurate. The communication guidelines determine the official spokesperson(s) and should proactive prepare templates for crisis communication to minimize confusion in critical situations, and maintain trust through communication. These simple guidelines promote transparency, and keep trustability to assigned team members.

Furthermore, the BCP core component focus on regular testing and training, because the business landscape is always changing as well as the risks and challenges that come with it. Therefore, the BCP should be actual, better advanced, and effective, which can be achieved by consistently putting the BCP through regularly tests and simulations to identify vulnerabilities, gaps and areas of improvement. Training and testing include developing a test methodology, simultaneous testing and training of the disaster recovery team, followed by BCP revision and simultaneous testing and training again. As a major component of the BCP, testing is essential to determine whether the BCP is adequate to address critical risks [18]. In addition to ensuring that the disaster recovery team members know what to do, testing under realistic conditions enabling develop confidence and avoid panic during a disaster event and stay ahead to ensure everyone can execute his part of the BCP when needed.

4.7 Emergency Communication Plan

Emergency Communication (ECO) and Emergency Communication Plan (ECOP) after NIS2 Article 21.2.j, is an essential component of the Disaster Recovery Plan (DRP) for Business Continuity (BC) and Cyber Resilience Planning (CRP). Detailed information on what should be protected and how internal and external emergency communication should be handled is an essential requirement that need the following aspects to be answered, such as:

- What must be protected in a disaster event?
- Identify potentially affected facilities/assets, important for ECOP,
- DRP contains measures on how operations should be kept running during a disaster event, how everything can be protected from hazards and when the system/devices should be shut down if the situation requires for,
- In addition, an ECOP must be available that leads to a plan "B" in case of a disaster event or failure.

4.7 Emergency Communication Plan

Regardless of whether protecting physical assets or limiting disruptions in digital operations is a priority, globally operating organizations must plan for:

Communication plan:

- Develop communication plan for internal and external stakeholders,
- Implement central communication nodes for real-time updates,
- Ensure redundancy in communication channels for reliability,

Continuous improvement:

- Update and improve cybersecurity measures,
- Conduct follow-up audits to learn from past incidents,
- Participate in knowledge exchange with associations/institutions/databases, expert panels, and others,

Critical assets:

- Security measures for important devices/systems, IT infrastructure, ... m
- Raise critical assets to a higher maturity level (risk assessment),
- Implement threat intelligence, which is evidence-based information about cybersecurity threat and cyber-attack events that cybersecurity experts organize and analyze, and may include attack mechanisms,

Data backup and recovery (DRP):

- Implement robust data backup and recovery systems,
- Store critical data at secure, off-site locations,
- Test data recovery systems regularly

Emergency Communication Plan (ECOP):

- Develop emergency response plan,
- Consider RPO, RTO, MTD,
- Conduct regular emergency drills to evaluate response behavior amd response times,

Employee training:

- Qualify employees in cybersecurity awareness and security measures,
- Regular training on cyberattacks, tactics, tools, and others,

Risk assessment:

- Identify assets and facilities a.t risk,
- Evaluate historical data and potential impacts,
- Work with IRT for accurate forecasting.

Supply chain resilience:

- Assess and strengthen supply chain cybersecurity,
- Identify security standards used in supply chain,
- Identify suppliers with critical security culture.

4.7.1 Important to Do's for ECOP: A Roadmap

Be ready to swiftly inform management of the situation, inform the designated company spokesperson of the situation

- Prepare a well-organized, step-by-step plan with relevant information, immediately available for company statements to traditional media, social media and other relevant institutions, organize and publish reporting,
- Prepare and use email, mobile phones, smartphones, social media platforms for dissemination of messages,
- Communicate situational information and procedures to employees and other stakeholders using the methods mentioned,
- Communicate with employees' families and local communities using the methods mentioned,
- Continually adapt technologies, message content and time frames for message delivery to changing events related to the emergency,
- Prepare to use social media to ensure that messages are appropriate to the situation and that only authorized employees are allowed to disseminate messages via social media.

Step-by-step Emergency Communication Plan (ECOP)

- List of internal contacts, e.g. employees,
- List of external contacts, e.g. traditional media, social media, suppliers, government agencies,
- Special forms, e.g. call logs to capture media and personal inquiries, emergency contact directory, report describing the incident,
- Pre-written documents, e.g. press releases, initial announcements and follow-up statements,
- Establish guidelines for dealing with social media in an emergency, e.g. who is allowed to issue messages, which media, social messaging platforms can be used,
- Trained emergency communications team with knowledge of messaging systems/social media,
- Trained company spokesperson and at least one deputy Technology to quickly disseminate emergency information to employees, stakeholders, suppliers, customers, government agencies and other external parties,
- Organization wide policies on all aspects of emergency communications.

Be prepared to communicate the following information in the event of an emergency

- Updated status reports on the incident,
- List of internal people contacted,
- List of external organizations contacted,
- Actions taken during the incident,
- Use and performance of electronic notification and messaging systems, especially social media,

4.7 Emergency Communication Plan

- Update of the ECOP based on experience with the event, problems encountered and how they were resolved,
- Ongoing problems requiring additional assistance: Description of the incident: what happened, what was done, the results and the consequences.

Prepare final status reports on the incident, presented in an after-the-event report

- List of internal people contacted,
- List of external organizations contacted,
- List of resources needed, acquired, used,
- Report on the performance of emergency reporting systems and social media,
- Actions taken to end the incident,
- List of problems encountered and their resolution,
- Report on the incident: what happened, what was done, the results and the consequences,
- Update ECOP based on lessons learned from the event.

4.7.2 ECOP Topics BCM, RPO, RTO, MtD

Managers (C-level) be responsible for BCM and must define tolerable times for optimized data recovery after cybersecurity threat and/or cyber-attack event emergency, using a risk-management strategy with regard to:

- Maximum tolerable recovery point (RPO),
- Maximum tolerable recovery time (RTO),
- Maximum tolerable downtime (MtD).

RPO, RTO, MtD represent important thresholds for mapping data security, disaster recovery and business continuity activities,

- Define and Quantify Resilience Value: Identify key metrics for effective measurement,
- Conduct Resilience Gap Analysis: Strategies for assessing current resilience and identifying potential for improvement,
- Quantifiable Business Scenario: Techniques for formulating Return on Investment (ROI) and converting resilience programs into strategic assets.#

4.7.3 Summarizing ECOP Action Needs

- Proactive measures to implement the obligations in the NIS2 directive!
- Ensure compliance with regulations/standards e.g. ISO/IEC27001, NIST-CSF, and others!
- Proactive training measures in cybersecurity awareness!

- Proactive measures in cyber hygiene!
- Usage of established IT security tools!
- Organization-wide and/or cross-organization implementation!
- Proactive security measures in an emergency situation are crucial (response) for short recovery times (RTO, RPO, MtD)!
- Fast recovery of BC after cybersecurity threat and/or cyber-attack events on essential and important facilities/entities must be a standard to do. Negative examples: insufficient budgets, outdated technologies, understaffed teams, and others!
- Necessary measures: Introduction of preventive protective measures, e.g. regular security audits, zero-trust architecture, regular cybersecurity training (cybersecurity awareness), ..., tailored to the individual requirements of the organization!
- Societies are dependent on essential and important facilities/entities, requires effective and efficient security measures not only for own protection, but also for protection of everyone!

4.8 Exercises

- What is meant by the term *RCE Directive?*
  - Describe the characteristics and capabilities of The RCE Directive
- What is meant by the term *Security Operations Center (SOC)*?
  - Describe the characteristics and capabilities of SOCs
- What is meant by the term *Pegasus*?
  - Describe the efforts and potential application domains of Pegasus
- What is meant by the term *Compliance?*
  - Describe the characteristics and capabilities of Compliance
- What is meant by the term *Liability?*
  - Describe the characteristics and capabilities of Lability
- What is meant by the term *Mandatory Cybersecurity Measures*?
  - Describe the characteristics and capabilities of Mandatory Cybersecurity Measures
- What is meant by the term *Standards in Cybersecurity Risk-Management*?
  - Describe the efforts and potential application domains of at least 10 Standards
- What is meant by the term *CMLM*?
  - Describe the characteristics and capabilities of CMLM
- What is meant by the term *Preparing for NIS2*?
  - Describe the efforts and potential action of at least 5 actions
- What is meant by the term *Vulnerability Management?*
  - Describe the characteristics and role of pen-testing

- What is meant by the term *Supply-chain Security Management*?
  - Describe the characteristics and capabilities in the context of internal and external risks
- What is meant by the term *Business Continuity Plan*?
  - Describe the efforts and potential actions of a Business Continuity Plan
- What is meant by the term *Likelihood*?
  - Describe the characteristics and capabilities of Likelihood
- What is meant by the term Component *Backup Schedule*?
  - Describe the characteristics set up in line with the Backup Schedule
- What is meant by the term *Recovery Point Objective (RPO)*?
  - Describe the characteristics, capabilities, and objectives of the Recovery Point Objective (RPO)
- What is meant by the term *Recovery Time Objective (RTO)*?
  - Describe the characteristics, capabilities, and objectives of the Recovery Time Objective (RTO)
- What is meant by the term *Standards in Cybersecurity Risk-Management*?
  - Describe the efforts and potential application domains of at least 10 Standards
- What is meant by the term *Maximum tolerable Downtime (MtD)*?
  - Describe the characteristics and terms of the MtD Equation
- What is meant by the term *Emergency Communication Plan*?
  - Describe the efforts and potential action of at least 7 actions to be taken

References

1. Directive (EU) 2022/2557 of the European Parliament and of the Council of 14 December 2022 on the resilience of critical entities and repealing Council Directive 2008/114/EC. https://eur-lex.europa.eu/legal-content/EN/TXT/PDF/?uri=CELEX:32022L2557 (accessed 01.2023)
2. Directive (EU) 2022/2555 of the European Parliament and of the Council of 14 December 2022 on measures for a high common level of cybersecurity across the Union, amending Regulation (EU) No 910/2014 and Directive (EU) 2018/1972, and repealing Directive (EU) 2016/1148 (NIS 2 Directive). ELI: http://data.europa.eu/eli/dir/2022/2555/oj (accessed 01.2023)
3. Negreiro, M.: ENISA and a New Cybersecurity Act. EU Parliament, 2019
4. Negreiro, M.: The NIS2 Directive – A High Common Level of Cybersecurity in the EU. Briefing EU Legislation in Progress, 2022
5. Pegasus Spyware. Wikipedia, 2024. https://en.wikipedia.org/wiki/Pegasus_(spyware) (accessed 10.2024)
6. Marczak, B., Scott-Railton, J., Razzak, B.A., Al-Jizawi, N., Anstis, S., Berdan, K., Deibert, R.: Pegasus vs. Predator: Dissident's Doubly-Infected iPohnes Reveals Cytrox Mercenary Spyware. Citizen Lab Research Report No. 147, University of Toronto, December 2021
7. Scott-Railton, J., Marczak, B., Razzak, B.A., Anstis, S., Herrero, P.N.-, Deibert, R.: New Pegasus Spyware Abuses Identified in Mexico. Citizen Lab Report No. 159, University of Toronto, October 2022.

8. Parsons, C.: Cybersecurity Will Not Thrive in Darkness: A Critical Analysis of Proposed Amendments in Bill C-26 to the Telecommunications Act, Citizen Lab Report No. 160, University of Toronto, October 2022
9. Rhopek, L.: Journalist Sues Predator Spyware Maker for Allegedly Helping Government Surveil Him, 2022. https://gizmodo.com/thanasis-koukakis-sues-intellexa-over-predator-spyware-1849625793 (accessed 01.2023)
10. Read the Indexa Pitch on its Spyware Tool. The New York Times, 2022. https://www.nytimes.com/interactive/2022/12/08/us/politics/intellexa-commercial-proposal.html (accessed 01.2023)
11. Federal Cybersecurity Research and Development Strategic Plan. Executive Office of the President of the United States, 2019
12. Federal Cybersecurity Research and Development Strategic Plan. Executive Office of the President of the United States, 2016
13. FY2022 Federal Cybersecurity R&D Strategic Plan Implementation Roadmap. A Report by the Cybersecurity & Information Assurance Interagency Working Group and Subcommittee on Networking and Information Technology Research, and Development Committee on Science and Technology Enterprise of the National Science and Technology Council. https://www.nitrd.gov/pubs/FY2022-Cybersecurity-RD-Roadmap.pdf (accessed 04.2023)
14. Draft of a law to implement the NIS 2 Directive and to regulate essential principles of information security management in the federal administration (in German), 2024. https://www.bmi.bund.de/SharedDocs/gesetzgebungsverfahren/DE/Downloads/kabinettsfassung/CI1/nis2-regierungsentwurf.pdf?__blob=publicationFile&v=2 (accessed 07.2024)
15. Savage, M.: Business Continuity Planning. In: Work Study, Vol. 51, No 5, pp. 254-261, 2018. https://doi.org/10.1108/00438020210437277 (accessed 03.2023)
16. Möller, D. P.F.: Guide to Cybersecurity in Digital Transformation – Trends, Methods, Technologies, Applications and Best Practices. Springer Nature, 2023
17. What is a Recovery Point Objective and How to Calculate One. In: Indeed Editorial Team, 2023
18. Cerullo, V., M. J.: Business Continuity Planning: A Comprehensive Approach. In: ISM Journal, pp. 70-78, 2004
19. Merritt, M., Hansche, S., Ellis, B., Sanchez-Cherry, K., Snyder, J. Walden, D.: Building a Cybersecurity and Privacy Learning Program. NIST SP 800-50 Rev. 1 (Initial Public Draft), 2023. https://csrc.nist.gov/pubs/sp/800/50/r1/ipd (accessed 12.2023)
20. Recovery Point Objective: Recovery Point Objective Definition, 2024. https://www.druva.com/glossary/what-is-a-recovery-point-objective-definition-and-related-faqs (accessed 08.2024)
21. Houghton, S.: Recovery Time Objective Explained, 2023. https://www.aztechit.co.uk/blog/recovery-time-objective (accessed 11.2023)
22. Maximum Tolerable Downtime (MTD) and Maximum Tolerable Data Loss (MTDL): Differences and Considerations. https://www.zerto.com/resources/a-to-zerto/mtd-and-mtdl-differences-and-considerations/ (accessed 01.2023)
23. True Cost of Cybersecurity. Field Effect 2024

Glossary

ACOR	Analytical Construction of Optimal Regulation: Mathematical problem considered in the general case for stabilizable systems of fractional order linear differential equations, whereby the regulation can be that the investigated system, e.g. a closed-loop feedback system, is asymptotically stable.
AD	Active Directory: Stores data as objects. Objects are defined as either resource, such as computers or printers or cybersecurity principals user groups.
AI	Artificial Intelligence: Theory and development of computer systems to make them able to perform tasks normally requiring human intelligence.
AMaaS	Additive Manufacturing-as-a-Service: Enable organizations to operate as contractors on demand, creating products for their business that employ them while distributing costs for software, equipment, maintenance and e
ANN	Artificial Neural Network: Consist of input, hidden, and output layers with connected neurons (nodes) to simulate the human brain.
APT	Advanced Persistent Threat: Cyber threat risk incident campaign in which a cybercriminal attacker, or team of cybercriminal attackers, establishes an illicit, long-term presence on a network in order to mine highly sensitive data.
ATT&CK	Adversarial Tactics, Technics & Common Knowledge: Is a set of data metrices, and assessment tools, developed by MITRE Corporation, to enable organizations understand their cybersecurity readiness and uncover vulnerabilities in their defenses.
B2B	Business-to-Business: Electronic commerce that focus on building long-term relationships to boost business growth, based on personal emotional relationship on understanding all the parties involved such that their roles in the business are fortified enough to last long.

B2C	Business-to-Customer: Service efforts targeted directly at customers and businesses try to forge an emotional bond with their present and future customers.
BC	Business Continuity: Organizations ability to maintain or swiftly resume acceptable levels of product or service delivery following a cyber threat risk incident that disrupts regular (normal) operations.
BCM	Business Continuity Management: Proactive approach that enable organizations to respond and recover from potential cyber threat risk incidents.
BCP	Business Continuity Plan: Document that describe how an industrial organization continues to operate in the event of an unplanned business interruption.
BIA	Business Impact Analysis: Systematic process to determine and evaluate the potential effects of an interruption to critical business operations as a result of a disaster, accident or emergency case.
BM	Backup Management: Ensure that less or no data are lost during a disaster through data backed up in a different cybersecure storage facility
BPM	Business Process Management: Methodology for managing the processes in an organization.
CAR	Causal Analysis and Resolution: Identify causes of selected outcomes and act to improve process performance.
CBA	Cost-Benefit Analysis: Data-drive approach evaluating a project or decision s and costs from a business perspective benefit, which mean forecasting profitability through a CBA to avoid financial loss.
CBW	Calculate Backup Window: Calculating time it takes to complete a backup from start to finish.
CCM	Cybersecurity Compliance Management: Aligning organizations cybersecurity processes with legal regulations and industry standards to protect digital data from emerging cyber threat risk incidents.
CEO	Chief Executive Officer: Highest-ranking executive in a organization.
CER	Critical Entities Resilience: Ability to prevent, protect against, respond to, resist, mitigate, absorb, accommodate and recover from cyber threat risk incidents that have potential to disrupt the provision of essential services.
CERT	Computer Emergency Response Team: Group of information security experts responsible for the protection against, detection of, and response to organizations cyber threat risk incidents.
CFA	Confirmatory Factor Analysis: Statistical technique used to verify the factor structure of a set of observed variables.
CI	Cyber Intelligence, also called Threat Intelligence: Evidence-based information about cyber threat risk incidents that cybersecurity experts analyze and organize. CI includes the types of strategic, tactical, and operational intelligence.
CIA Triad	Confidentiality, Integrity, and Availability: Common model that forms the basis for the development of cybersecurity systems. Also used for finding vulnerabilities and methods for creating cybersecurity solutions.

CIO	Chief Information Officer: Highly technical person that oversee IT-departments resources and staff.
CIP	Continuous Improvement Process: Ongoing and incremental cycle or process improvement to identify, define, implement, measure, and analyze the effectiveness of a process change.
CISA	Cybersecurity & Infrastructure Security Agency: Mission is to defend and cybersecure cyberspace by leading national efforts to drive and enable effective national cyber defense, resilience of national critical functions, and a robust technology ecosystem.
CISCSC	Center for Internet Security Critical Security Control: Result of input from cybersecurity experts around the world to better defend against known cyber threat risk incidents by distilling key cybersecurity concepts into actionable controls to achieve greater overall cybersecurity defense.
CISO	Chief Information Security Officer: Senior-level executive that oversees an organizations information, cyber and technology security.
CM	Crisis Management: Enables organizational and/or operational continuity.
CML	Cybersecurity Maturity Level: Defines the readiness level to defend against cyber threat risk incidents.
CMLM	Cybersecurity Maturity Level Model: Measures the cybersecurity capabilities over time and identify target maturity levels based on cyber risk, and prioritize actions that enable them to achieve the targeted level.
CMM	Capability Maturity Model: Methodology applied to develop and refine organizations software development processes.
CMMI	Capability Maturity Model Integration: incorporates best components of individual disciplines of CMM like Software CMM, Systems Engineering CMM, and other.
CMMI-DEV	CMMI for Development: Integrated set of best practices that improves performance and key capabilities for organizations that develop and maintain products, components, and services for digital system development.
CMMI-DEV+IPPD	CMMI-DEV +Integration of Product and Process Development: Integrated set of best practices that improves performance and key capabilities for organizations that develop and maintain products, components, and services for digital system development that simultaneously integrates all activities through the use of IPPD to optimize design, manufacturing, and supportability processes. Hence, IPPD facilitates meeting cost and performance objectives from product concept through production, including field support, which meet efforts like Concurrent Engineering to improve customer satisfaction and competitiveness in a global economy.

CMMI-SVC	CMMI for Service: Integrated set of best practices that improves performance and key capabilities for organizations that provide services, including B2B, B2C, as well as standalone services, and those that are part of a product offering.
CMS	Compliance Management System: Processing of monitoring organizations systems, policies, and procedures to ensure all personal levels, from CEO, CIO, ..., employees, comply with federal, state and local laws, governmental directives and regulations, accreditation rules, and codes of conducts.
CNO	Computer Network Operations: Involve the use of several tools and techniques such as reconnaissance, acces, exfiltration and system compromise.
COBIT	Control Objectives for Information and Related Technologies: Enables organization to meet business challenges in in regulatory compliance, risk-management, and aligning IT strategy with organizational goals.
COSO	Committee of Sponsoring Organizations Framework: Demonstrate in a 3D diagram how all elements of an internal control system are related.
COTS	Commercial Off-The-Shelf: Software or hardware product commercially ready-made and available for sale, lease. Or license to the general public.
CP	Contingency Planning: Preparing organizations being ready to respond effectively in the event of a cyber risk.
CPE	Common Platform Enumeration: Standard machine-readable format for encoding names of IT products and platforms.
CPS	Cyber-Physical Systems: Integrate sensing, computation, control and networking into physical objects and infrastructure, and connecting them to the Internet and to each other.
CRA	Cybersecurity Risk Assessment: Ongoing ability to protect organizations data, assets and network and information systems from cyber threat risk incidents.
CRITIS	Critical Infrastructure Systems: Systems, facilities and digital assets that are vital for operating industry, business, government, and society.
CRM	Cybersecurity Risk-Management: Process of identifying organizations digital assets, ongoing reviewing cybersecurity measures and implementing solutions, to either continue with what effective works or to mitigate cyber risks that may pose cyber threat risk incidents to as business.
CSA	Cybersecurity Situational Awareness: Ongoing process on understanding the cyber threat risk incidents environment within which it operates, its associated cyber risk impacts, and the adequacy of its risk mitigation measures.

CSIRT	Computer Security Incident Response Team: Assist in responding to computer security-related incidents.
CTEM	Cyber Threat Exposure Management: Practice of managing potential cyber threat risk incidents to corporate cybersecurity. It involves cyber risk identification, prioritization, and management across organization´s digital attack surface.
CTI	Cyber Threat Intelligence: Any information that enable an organization identifying, assessing, monitoring, and responding to cyber threat risk incidents.
CVE	Common Vulnerabilities and Exposures: Method for publicly sharing information on cybersecurity vulnerabilities and exposures.
CWE	Common Weakness Enumeration: Aims to provide a common base to identify the type of software weakness (vulnerability).
DCS	Distributed Control System: Computerized control system for a process with many control loops, whereby automated controllers are distributed throughout the system without no central operator supervisory control.
DDoS	Distributed Denial of Service: Cybercrime in which the attacker floods a server with Internet traffic to prevent users form accessing connected online services and sites.
DL	Deep Learning: Type of Machine Learning (ML) based on Artificial Neural Networks (ANN) in which multiple processing layers used to extract higher level features.
DMA	Digital Market Act: Set of clearly defined objective criteria to qualify large online platforms as a so-called gatekeeper to ensure that they behave in a fair way online and leave room for contestability, whereby obligations must comply within their daily operations.
DNS	Domain Name System: Turns domain names into IP addresses, which browsers use to load Internet pages.
DORA	Digital Operational Resilience Act: Address critical gaps in EU financial regulation. Thus, financial institutions are required to follow stringent guidelines for safeguarding against ICT-related cyber threat risk incidents.
DoS	Denial of Service: Cyber threat risk incident overload a website or network with the aim of degrading its performance or even make it completely inaccessible.
DR	Disaster Recovery: Enable technological facility continuity.
DRP	Disaster Recovery Plan: Documented, structured approach that describes a step by step approach how an organization can swiftly resume work after an unplanned cyber threat risk incident.
DSA	Digital Service Act: Create a safer digital space where fundamental rights of users are protected. It includes a large category of online services, from simple websites to internet infrastructure services and online platforms that must abide by strict obligations of the Act.
ECOP	Emergency Communication Plan: Document that provides guidelines, contact information and procedures for how information should be shared during all phases of an unexpected occurrence that requires immediate action.

EDRT	Estimate Data Restoration Time: Measures the time from detection of a cyber threat risk incident or outage until the full system functionality is restored.
EFA	Exploratory Factor Analysis: Used to discover the factor structure of a measure to examine its internal reliability.
ENISA	European Union Agency for Network and Information Security: Centre of expertise for cybersecurity in Continental Europe European Union that support the EU and EU-Member States to be better equipped and prepared to prevent, detect, and respond to information security problems.
ETSI	European Telecommunications Standards Institute: Independent, not-for-profit standardization organization operating in the field of information and communication
EU-CyCLONe	European Cyber Crisis liaison organization network: Cooperation network of EU-Member States national authorities in charge of cyber crises management.
FAIR	Factor Analysis of Information Risk: Quantify cyber risks and define the changes of them becoming serious cyber threat risk incidents.
GenAI	Subset of AI that creates new content based on learned patterns and rules, such as text, images, audio, and code.
GPO	Group Policy Object settings: Identify and mitigate cyber threat risk incident paths that cybercriminals could not do to seize control of the domain, detect and respond to Indicators of Compromise (IoC), and be prepared to swiftly restore AD domain or forest.
GPT	Generative Pretrained Transform: Family of neural network models that uses a transformer architecture to create something new, representing an enhancement in AI. GPT models enable applications to create human-like text documents and content, as well as answer questions in a conversational form.
GRDP	General Regulation Data Protection: List the right of data subjects, which means the right of individuals whose personal data is being processed.
HMI	Human Machine Interface: Feature or component of a certain device or software application, enabling humans to engage and interact with machines.
I4.0	Industry 4.0: Refers to the fourth industrial revolution, is the cyber-physical transformation of manufacturing.
IaaS	Infrastructure-as-a-Service: On-demand availability of highly scalable computing resources as services over the internet.
IACS	Industrial Automation and Control System: Refers to the integration of devices, machines, and equipment within a manufacturing environment like Industry 4.0.
IAM	Identity and Access Management: Enable that only the right user can access organizations data, digital assets and resources.

Glossary

ICIF	Internal Control Integrated Framework: Structure that integrates various internal control concepts into fife interrelated components: control environment, risk assessment, control activities, information and communication, and monitoring, to enable managers better control their organization.
ICS	Industrial Control System: Consists of combinations of control components (e.g., electrical, mechanical, hydraulic, pneumatic) that act together to enable to better industrial objective (e.g., manufacturing, and others).
ICSR	Individual Case Safety Reports: Report to regulatory authorities in accordance with strict requirements and timelines
ICT	Information and Communication Technology: Covers all technical means used to handle information and communication.
IDS	Intrusion Detection System: Detect suspicious activities and generates alerts when detected.
IDSA	Intrusion Detection System Architecture (IDSA): Machine Learning (ML)-based method to detect data, patterns or signatures of potential cybersecurity breaches through cyber threat risk incidents.
IEC	International Electrotechnical Commission: Worldwide used IEC International Standards and Conformity Assessment systems to ensure safety.
IED	Intelligent Electronic Devices: Device that incorporate one or more processors with the capability to receive or send data or control from or to an external source.
IIoT	Industrial Internet of Things: Consists of sensors, instruments, machines, and other devices that are networked together and use internet connectivity to enhance industrial and manufacturing business processes and applications. IIoT includes Machine Learning (ML) and Big Data Technology along with Programmable Logic Controller (PLC), SCADA, Remote Terminal Unit (RTU), Human Machine Interface (HMI), Intelligent Electronic Devices (IED).
IoC	Indicators of Compromise: Refers to data that indicates that a system has been infiltrated by a cyber threat risk incident.
IotF	Industries of the Future: Specifically are Quantum Information Science (QIS), Information Technology, Artificial Intelligence (AI), Autonomous Systems, Advanced Manufacturing, Automated Delivery Systems, Robotic Surgery and Diagnostics, 6G, Biotechnology, and others.
IRT	Incident Response Time: Speed at which an organization can identify, respond to, and mitigate a cyber treat risk incident.
ISACA	Information Systems Audit and Control Systems: International professional association that focus on IT governance.
ISMS	Information Security Management System: Set of policies and procedures for systematically managing organizations sensitive data.
ISO	International Organization for Standardization: Worldwide federation of national standards bodies.

IT	Information Technology: Use of any computers, storage, networking and other physical devices, digital resources and processes to create, process, store, secure and exchange all forms of electronic data.
KSA	Knowledge, Skills, and Abilities: Task statements required as building blocks or the NIST NICE framework.
LEF	Loss Event Frequency: Represent a timeframe, a cybercriminal can inflict harm on an asset.
LEM	Loss Event Magnitude: Represents the severity, i.e., scope and impact of a cyber threat risk incident.
LLM	Large Language Model: Type of Artificial Intelligence (AI) program, build on Machine Learning (ML) that can recognize and generate text, among other tasks, trained on huge sets of data.
MA	Mission Assurance: Process to protect and ensure the continued function and resilience of capabilities and assets be refining, integrating, and synchronizing the aspects of DoD security, protection, and risk-management programs that directly relate to mission execurion.
MFA	Multi-Factor Authentication: Multi-step account login process that requires users to provide two or more verification factors to gain access to an organizations resource. MFA is a core component of a strong Identity and Access Management (IAM) policy.
MITRE	Massachusetts Institute of Technology Research and Engineering. Government-funded research organization that has beside others a substantial cybersecurity practice funded by NIST.
ML	Machine Learning: Based on Artificial Neural Networks (ANN) in which multiple layers of processing are used to extract higher level features from data.
MtD	Maximum tolerable Downtime: Refers to the maximum duration a system can remain inoperable before causing severe impact on an organization.
NICE	National Initiative for Cybersecurity Education: Foundation for increasing the size and capability of the U.S. cybersecurity workforce. NICE provide a common definition of cybersecurity, a comprehensive list of cybersecurity tasks, and the knowledge, skills, and abilities (KSAs) required to perform those tasks, which provides organizations with a systematic approach to analyze and assess their business models.
NIS	Network and Information Security: First EU-wide cybersecurity legislation.
NIS2	Network and Information Security 2: Stringent regulation that requires EU-wide vital sectors to improve their cybersecurity measures, incident response plan, and reporting mechanisms.
NIST	National Institute of Standards and Technology: Promotes U.S. innovation and industrial competitiveness by advancing measurement science, standards, and technology in ways that enhance economic security and improve quality of life.

Glossary

NIST CSF	U.S. National Institute of Standards and Technology Cybersecurity Framework: Flexible set of guidelines and best practices developed to enable organizations enhance their information security and manage cybersecurity risks. It consists of standards, guidelines and best practices to manage cybersecurity risk.
NIST CSF 2.0	National Institute of Standards and Technology Cybersecurity Framework v2.0: Updated version of NIST CSF, which introduce updates that enhance its focus on governance by adding a sixth core function called govern, supply chain risk-management and measurement of cybersecurity outcomes, aligning more closely with international standards and improving flexibility for organizations.
NIST SSDF	National Institute of Standards and Technology Secure Software Development Framework: Set of fundamental, sound, and cybersecure software development practices based on established cybersecure software development practice documents from organizations such as BSA, OWASP, and SAFECode.
NLP	Natural Language Processing: Machine Learning (ML) technology that enables computer systems to interpret, manipulate, and comprehend human language.
OCTAVE	Operationally Critical Threat, Asset and Vulnerability Evaluation: Focusses on assessing organizational cyber risks, rather than technological risks.
OPM	Organizational Performance Management: Efficiently managing and improving the performance of operational processes within an organization that is crucial for enhancing productivity, efficiency, and overall success.
ORT	Operational Resumption Time: Time that begin to run afresh for the purposes of the time limits.
OS	Operating System: Refers to the program that, after being loaded into the computer by a boot program, manages the other application programs in a computer.
OSINT	Open-Source INTelligence: Framework that provides a plethora of details for open information sources.
OT	Operational Technology: Technology that interfaces detects with the physical world and includes Industrial Control Systems (ICS), Supervisory Control and Data Acquisition (SCADA) and Distributed Control Systems (DCS) for direct interaction (monitoring, control and or change) of industrial equipment, assets, processes and events.
OTA	Over the Air: Any means of broadcast transmission
OTP	One-Time Password: Use a special algorithm to generate a unique one-time password. Algorithm uses the current time and a secret key that is shared between a user's device and the requested ones.
OTSEC	Operational Technology Security: Protect and control critical digital systems such as infrastructure systems.

OWASP	Open Worldwide Application Security Project: Nonprofit foundation dedicated to improving software cybersecurity supporting website owners and cybersecurity experts protect web applications from cyber threat risk incident attacks.
OWASP CLASP	Open Worldwide Application Security Project Comprehensive Lightweight Application Security Process: Activity-driven, role-based set of processes to build cybersecurity into an existing or new software development program in a structured, repeatable and, measurable way.
OWASP SAMM	Open Worldwide Application Security Project Software Assurance Maturity Model: Mission is to provide an effective and measurable way to analyze and improve a cybersecure development lifecycle.
PaaS	Platform-as-a-Service: Cloud computing service model over the Internet that offers a flexible, scalable cloud platform to develop, deploy, run, and manage apps.
PACS	Physical Access Control System: Type of physical cybersecurity system designed to control access to an area like who granted access, when the access is granted, and how long the access should last.
PADM	Policy And Documentation Management: Determines how documents are created, reviewed, managed, and stored within organizations, to ensure compliance, accuracy, and efficiency.
PAM	Privileged Access Management: Identity cybersecurity solution enabling to organizations against cyber threat risk incidents by monitoring, detecting, and preventing unauthorized privileged access to critical resources.
PE	Processing Element: Component of an Artificial Neural Network (ANN) equivalent to neurons that performs mappings of input values based on those inputs, either generate an (single) output value. PEs are connected together to form a network.
PLC	Programmable Logic Controller: Used in SCADA and DCS systems as control component of an overall hierarchical system to locally manage processes through feedback control.
POA&M	Plan of Action & Milestones: Corrective plan to address system weaknesses and the resources required to fix them.
PTaaS	Pentest-as-a-Service: Aligns with modern cybersecurity demands through a SaaS-based model, faster start times, and the power of a diverse, vetted cybersecurity community.
QIS	Quantum Information Science: Capabilities that harness quantum mechanics and quantum material properties to achieve computation, information processing, communications, and sensing in ways that cannot be achieved by classical physics principles.

QMS	Quality Management System: Formalized system that documents processes, procedures, and responsibilities for achieving quality policies and objectives.
RaaS	Ransomware-as-a-Service: Malware offer that can be used for a fee.
RAF	Risk Assessment Framework: Structured set of activities to evaluate and manage cyber threat risk incidents in a systematic manner.
RCE	Resilience of Critical Entities: Lays down obligations of EU Member States, to ensure that essential services for the maintenance of vital societal functions or economic activities are provides in an unobstructed manner within the internal EU market.
RDP	Remote Desktop Protocol: Secure network communication protocol enable users to execute remote operations on other computers.
RFID	Radio Frequency Identification: Wireless communication form that incorporates the use of electromagnetic or electrostatic coupling in the radio frequency portion of the electromagnetic spectrum to identify an object.
RMF	Risk-Management Framework: Set of practices, processes, and technologies to identify, assess, and analyze cyber risk to manage cyber threat risk incidents within an organization.
RPO	Recovery Point Objective: Defines the maximum amount of data, measured by time, which can be lost after a recovery from a disaster, failure, or comparable event before data loss will exceed what is acceptable to an organization.
RTO	Recovery Time Objective: Maximum tolerable length of time that a computer, system, network or application can be down after a failure or disaster occurs.
RTOS	Real-Time Operating System: Operating System with the key features' predictability and determinism. RTOS are subdivided into soft real-time and hard real-time systems.
RTU	Remote Terminal Unit: Self-contained computer with various components including a processor, memory and storage.
SaaS	Software-as-a-Service: Allows users to connect to and use cloud-based apps over the Internet.
SbD	Secure by Design and Default: Approach to security system/product development where customer-centric security requirements included into system/product development from the beginning and products are released in a state where customers are secure out-of-the-box.
SBOM	Software Bill of Materials: Key building block in software security and software supply chain risk-management, and list of ingredients that make up the software.
SCADA	Supervisory Control and Data Acquisition: Category of software applications for controlling industrial processes, which is the gathering of data in real time from remote locations in order to control equipment and conditions.

SCAMPI	Standard CMMI Appraisal Method for Process Improvement: Method to provide benchmark-quality ratings relative to CMMI models to evaluate organizations' processes and provide ratings.
SCP	Secure, Contain, Protect: Refer to special containment procedures.
SCRM	Supply Chain Risk-Management: Process finding and addressing potential vulnerabilities in organizations supply chains.
SDLC	System Development Life Cycle: Project management model that encompasses system or software creation from its initial idea to its finalized deployment and maintenance.
SDP	Software-defined Perimeter: Cybersecurity methodology that distributes access to internal applications based on a user´s identity, with trust that adapts based on context.
SEM	Security Event Management: Process of monitoring, correlating and managing cybersecurity events within an organization IT infrastructure to detect and respond to potential cyber threat risk incidents.
SIEM	Security Information and Event Management: Cybersecurity approach that combines Security Information Management (SIM) and Security Event Management (SEM) functions into one cybersecurity management system.
SIM	Security Information Management: Collecting, monitoring, and analyzing cyber security related data from security logs and various other data sources.
SLA	Service Level Agreement: Contract between a service provider and its customers that documents what services the provider furnish and defines the service standards the provider is obligated to meet.
SML	Supervised Machine Learning: Category of Machine Learning (ML) that use labeled datasets to train the ML algorithm to predict outcomes that are either regular (normal) or abnormal in contrast to used training data.
SOC	Security Operations Center: Improves organizations cyber threat risk incident detection, response and prevention capabilities by unifying and coordinating all cybersecurity technologies and operations.
SQL	Structured Query Language: Standard language for database creation and manipulation.
SSDLC	Software Development Life Cycle: Defines cybersecurity requirements and tasks that must be considered and addressed within every system, project or application that is created or updated to address a business need.
SSE	Systems Security Engineering: Applies scientific, engineering, and information assurance principles to deliver trustworthy systems that satisfy requirements within established cyber risk tolerance.
SSH	Secure Socket Shell: Protocol for securely sending commands to a computer over an unsecured network.
SSI	Snapshot Interval: Select the desired interval to change the interval between snapshots.

SSL	Secure Socket Layer: Networking protocol designed for securing connections between web clients and web servers over an insecure network, such as the internet.
SSL	Secure Socket Layer: Standard technology for securing an Internet connection by encrypting data sent between a website and a browser or between two servers.
SSML	Semi-Supervised Machine Learning: Method combines the benefits of the SML and UML methods, and be applied to automate feature learning and work with unstructured data.
SSO	Single Sign On: Identification method that enables users to log in to multiple applications and websites with one set of credentials.
TARA	Threat Assessment and Remediation Analysis: Method to identify and assess cyber vulnerabilities and select countermeasures effective at mitigating those vulnerabilities.
TDDI	Trustworthy Distributed Digital Infrastructure: Technologies that facilitate cybersecure information communications infrastructure, enabling next-generation wireless communication, distributed computing, seamless integration of telecommunication systems with Cyber-Physical Systems (CPS), and provides the communications infrastructure for the Industries of the Future.
TFEU	Treaty on the Functioning of the European Union: Sets out organizational and functional details of the European Union, accompanied by protocols and declarations, as well as the charter or fundamental rights of the EU.
TI	Threat Intelligence: Evidence-based information about cyber threat risk incidents that cybersecurity teams organize and analyze.
TLD	Top Level Domain: Describes the part of a URL that represents the final section of a domain name.
TOTP	Time-based One-Time Password: Time-based OTP.
TSL	Transport Layer Security: Widely adopted security protocol designed to facilitate privacy and data security for communications over the Internet.
TtD	Time to Data: Describes the required time to retrieve the backup data and deliver it to the restore location.
TTD	Time To Detect: Performance indicator in cyber threat risk incident management.
TTP	Tactics, Techniques and Procedures: Term used by cybersecurity professionals to describe the behaviors, processes, actions, and strategies used by cybercriminal executing cyber threat risk incidents.
UML	Unsupervised Machine Learning: Category of Machine Learning (ML) that Learns from data without human supervision. UML can find patterns from input data and make assumptions about what data is perceived as regular (normal).
URL	Uniform Resource Location: Unique identifier used to locate a resource on the Internet
VDP	Vulnerability Disclosure Program: Showcase organizations priority to cybersecurity publicly.

XML Extensible Markup Language: Enables information exchange between computer systems like websites, databases, and third-party applications.

ZTNA Zero Trust Network Access: IT security solution that provides secure remote access to an organization's applications, data, and services based on clearly defined access control policies.

Index

A
Access risk-management, 210, 211
Adapt, 7, 9, 11, 17, 43, 51, 53, 54, 58, 66, 76, 97, 112, 119, 159, 163, 179, 205, 248
   to emerging cyber threat risk incidents, 51, 97
Additive Manufacturing
   as-a-Service, 3
Advanced
   competencies, 9
   Persistent Threats (APT), 53, 182
Analytical Construction or Optimal Regulation (ACOR), 194
Analytics, 4, 5, 27, 77, 182, 218
   cognitive, 4
   descriptive, 4
   diagnostic, 4
   predictive, 3–5
   prescriptive, 4, 5
Analyze incident response capabilities, 210
Anomaly
   detection, 61, 65–66
Anticipate, 10, 17, 53, 55, 172, 220
Anti-malware log files, 16
Appraisal Requirements for CMMI (ARC), 207, 208
Artificial
   algorithm, 3, 58
   -based cyber-attacks, 61
   Intelligence (AI), 3–4, 12, 19, 56, 57, 121, 153, 154, 188, 220
   Neural Networks (ANN), 12, 59
   powered monitoring, 61
   technology, 61

Aspects concerning supervision and enforcement, 141
Assesses, 19, 21, 23, 29, 47, 50, 56, 65, 90, 92, 94, 100, 102, 103, 109–112, 128, 138, 145, 151, 157, 160, 163, 164, 167, 174, 195, 196, 202, 204, 210, 219, 222, 223, 227, 228, 235, 243, 247
Assessments, 15–17, 19–21, 23, 28, 47, 48, 77, 79, 80, 83, 85, 88, 89, 91, 95–97, 99, 106, 108–110, 113, 119, 125, 128–132, 134–137, 151–170, 174, 175, 178, 181, 189–192, 194, 198, 208–212, 218, 223–225, 229, 235, 238, 242, 244, 247
Asset
   business, 9, 12, 14, 16, 116, 117, 152, 162, 163, 236
Attack
   memory-based, 117
   network-based, 117
   phishing, 18, 50, 61, 121
   ransomware, 45, 47, 121, 239, 245
   social-engineering, 53, 61, 211
   spear-phishing, 53, 181
   web application, 117
   Wi-Fi, 117
   XXS, 116
   zero-day angle, 118
Audit log management policy, 187
Authentication
   biometric, 124, 233
   knowledge-based, 123
   possession-based, 123
Authorizes, 31, 44, 161

Index

Availability
  requirements, 38, 39

B
Backup
  data, 24, 27, 44, 52, 242, 243, 247
  and disaster recovery, 26, 78, 94, 100–102, 111, 240
  frequency schedule, 24
  reliable, 24, 166, 240
Behavioral Detection Analytics (BAD), 181
Benchmark against industry standards, 211
Big data
  technology, 5, 36
Big Data and Analytics, 2, 4, 5, 12
Business
  continuity (BC), 15, 24, 53, 75–77, 92, 94, 96, 111, 129, 130, 151, 155, 165, 166, 169, 227, 229, 234–236, 240, 243, 246, 249
  Continuity Management (BCM), 30, 94, 100–102, 129, 134, 200, 224, 235, 249
  Continuity Plan (BCP), 26, 101, 102, 165, 167, 234, 236–246
  Impact Analysis (BIA), 26, 167, 177, 238, 243, 244
  Process Management (BPM), 7
  Resumption Plan, 101
  to-Business (B2B), 7, 204
  to-Customer (B2C), 7
  transformation, 8

C
Calculate
  Backup Window (CBW), 243
  data processing time, 243
Capability
  level, 202–208, 233, 235
  maturity level, 201–209
  Maturity Model (CMM), 202, 203
Capacity
  utilization, 3
Categorize, 102, 117, 160
Causal Analysis and Resolution (CAR), 207
Change management, 2, 38, 65
ChatGPT, 12, 60
CIA triad
  pillar availability, 47
  pillar confidentiality, 46
  pillar integrity, 47
CIS Controls
  v8.1, 186–188
Clarity, 186
Close loop system, 35
Cloud
  computing, 2, 5, 6, 12, 128, 133, 137, 141
  Disaster Recovery Plan (DRP), 25, 26, 47, 101, 111, 167, 169, 238, 239, 246
  services, 2, 5, 6, 111, 210, 238
Codecov Bash uploader, 103
Coexistence, 186
Collaboration culture, 2
Committee
  of Sponsoring Organizations of the Treadway Commission (COSO), 162
  procedure, 134, 144
Common
  Platform Enumeration (CPE), 31
  Vulnerabilities and Exposures (CVE), 31
  Weakness Enumeration (CWE), 31
Communication guidelines, 246
Community Cyber Security Maturity Model (CCSMM), 202
Competent authorities and single points of contact, 85
Compliance Management System (CMS), 169–170
Component lifetime in OT systems, 39
Compromising, 104, 114
Computer
  Network Operations (CNO), 66
  Security Incident Response Teams (CSIRTs), 82, 84–91, 131–133, 135, 140, 141, 145
Conditions for imposing administrative fines on essential and important entities, 142, 225
Confidentiality
  integrity, availability (CIA Triad), 55
Confirmatory Factor Analysis (CFA), 194, 195
Connectivity, 2, 12, 13, 15, 35, 36, 38, 81, 171
Containment, 47, 52, 98, 100, 139, 210
Contexts, 2, 7, 10, 12–15, 22, 26, 29, 31, 34, 37, 38, 40–44, 48, 50, 56, 58, 60–62, 80, 95, 101, 108, 123, 127, 128, 142, 146, 166, 172–175, 179, 180, 182, 183, 186, 189, 190, 195, 205, 217, 221, 223, 232, 233, 236, 242, 245
Contingency planning, 165–169, 212, 229
Continuity of operations plan, 101
Control objectives, 35, 162, 197
Cooperation
  group, 87–91, 127, 128, 130, 134, 145
  international, 90
  at national level, 85, 87
  at Union and International Level, 81, 221

Index

Coordinated
 Cybersecurity Frameworks, 81, 84–87, 221
 vulnerability disclosure and European vulnerabilities, 86
Cost
 effective security, 23
 predictable, 6
 reduced total, 6
Credential management policy, 187
Critical Entities Resilience (CER), 54, 76, 196
Cross-site scripting (XXS) attacks, 103, 116
CSIRTs network, 87–91, 133, 145
Cultural transformation, 7, 8
Cyber
 attacker engagement, 182
 attack landscape, 17
 breaches likelihood, 199
 Denial, 182, 183
 environment, 17
 hygiene policies, 152
 hygiene practices, 81, 94, 113–121, 154
 insurance, 121
 insurance providers, 22
 Intelligence (CI), 51
 resilience, 15, 19, 51–54, 75–77, 119, 152, 189, 218, 221, 231
 Resilience ACT (CRA), 127, 230
 risk analysis, 33, 48, 49, 94–98, 157, 224
 risk assessment, 15, 16, 19, 21, 48, 68, 91, 109, 110, 113, 128, 137, 155–156, 161–164, 175, 178, 198, 223, 225, 238
 risk documentation, 154
 risk evaluation, 20–21
 risk impact, 39
 risk identification, 19, 130, 154, 156–158, 211
 risk levels, 15, 21, 48–50, 160, 237
 risk likelihood, 79
 risk management, 15, 16, 19, 21–30, 45, 54, 68, 77, 81, 82, 84, 91–137, 147, 151, 152, 154, 155, 165, 172, 173, 175, 177–179, 196, 199, 200, 217, 226
 risk measurements, 49
 risk review, 54, 195, 218
 Threat Exposure Management (CTEM), 119
 Threat Intelligence (CTI), 50, 51, 158, 181
 threat risk exposure, 26
Cybercriminal
 advantage, 18
Cybersecurity, 111, 129, 140, 167, 185
 audit, 80, 110, 223, 225
 awareness, 10, 14, 15, 18, 27, 28, 30, 48, 51, 85, 91, 92, 94, 120, 125, 176, 197–200, 222, 232, 234, 236, 238, 247, 249, 250

breach, 24, 37, 45, 65, 126, 153, 210
Capability Maturity Model (CCMM, C2M2), 201–209
Capability Maturity Model Integration (CMMI), 202–209
Capability Maturity Model Integration for Development (CMMI-DEV), 202, 205, 208, 209
Capability Maturity Model Integration for Services (CMMI-SVC), 204
chain, 23
Compliance Management (CCM), 30
Enhancement Act, 55, 219
flaws, 16
gaps, 114, 171
information-sharing arrangements, 91, 139
and Infrastructure Security Agency (CISA), 31
maturity level (CML), 27, 30, 48, 96, 112, 114, 151, 158, 160, 161, 170, 171, 200, 201
Maturity Level Model (CMLM), 27–30, 197–201, 233
Maturity Model (CMM), 196–209
policies and procedures, 38, 51, 121–123, 145–146, 210
posture, 17, 19, 23, 27, 34, 46, 50, 52, 56, 78, 112, 113, 118, 121, 159, 165, 168, 170–172, 174, 209–211, 225
Research And Development Strategic Plan, 55, 56, 220
risk assessment (CRA), 15, 16, 19–21, 48, 91, 109, 163, 198
risk landscape, 109, 155, 218
risk management (CRM), 15, 16, 19, 21–30, 45, 154, 172, 173, 175, 178, 179, 199, 226–234
risk management measures, 84, 91–137, 152
risk management measures and reporting obligations, 82, 91–137
rules and controls, 109
SCAMPI Appraisal Method, 207–209
Service for the Union Institutions, Bodies, Offices and Agencies (CERT-EU), 127
situational, 16–19, 81, 97, 166
skills, 17, 19, 29
strategy, 15, 17, 22, 23, 26, 30, 38, 45, 48, 54, 82, 84, 85, 91, 156, 157, 173, 180, 221, 223
threat modeling awareness, 117
training, 94, 100, 113–121, 200, 210, 222, 224, 227, 234, 236, 250
weaknesses, 23, 79, 95
Cytrox, 219

D

Data
- analytics, 2–5, 13, 27, 196, 218
- availability, 242
- backup, 24, 27, 44, 52, 168, 169, 242, 243, 247
- base, 5, 12, 48, 60, 64, 103, 110, 122, 139, 209, 242, 243, 247
- base of domain name registration data, 139
- breach, 45, 46, 52, 121, 135, 142–143, 210, 224, 225, 227, 236, 245
- Center Disaster Recovery Plan (DRP), 247
- confidentiality, 39, 47, 166
- digital, 1, 5, 12, 16, 18–24, 26, 27, 35, 41, 50, 61, 65, 66, 77, 95, 96, 110, 111, 113–115, 121–124, 154, 158, 159, 164–166, 200, 221
- encryption, 44, 111, 121, 153
- integrity, 32, 44, 47, 169
- loss, 20, 24, 25, 44, 52, 101, 102, 159, 166, 168, 169, 171, 237–244
- machine readable, 1
- quality monitoring, 64
- recovery, 19, 156, 247

Deception technology, 183

Decision
- making, 3, 5, 13, 17, 35, 56, 58, 65, 90, 113, 158, 162, 237, 243
- proactive, 19
- tree analysis, 158

Deep learning (DL), 4, 12, 59, 60, 63, 65, 66
Defense capabilities, 158
Definitions, 21, 33, 58, 84, 107, 135, 158, 159, 168, 170, 186, 187, 189, 193, 208, 226
Delegated and Implementing Acts, 81, 144
Denial of Service (DoS), 41, 44, 53, 64
Detect (DE), 17, 28, 29, 35, 38, 41, 43, 47, 52–55, 57, 60–65, 77, 111, 139, 159, 174, 176, 178, 182, 183, 196, 199, 209, 218–221
Detection, 4, 6, 17, 23, 28, 44, 47, 48, 51, 60, 62, 64–66, 94, 104, 105, 110, 117, 139, 141, 153, 181, 195, 196, 217, 225, 234
Deter, 15, 55, 65, 220

Determine
- cybersecurity risk levels, 15
- RTO, 26, 243

Digital
- awareness, 11
- business transformation, 8
- ecosystem, 7, 8, 13
- evolution, 1
- landscape, 12, 75
- Market Act (DMA), 13
- model, 3
- Operational Resilience Act (DORA), 111, 119
- Service Act (DSA), 13, 14, 30, 199
- transformation, 1–15, 37, 40, 41, 44, 54, 76, 170, 192, 217, 218
- transformation paradigm, 2, 7

Digitalization, 1–66, 76, 81, 103, 217

Disaster
- recovery, 24–26, 78, 94, 100–102, 111, 167, 168, 227, 237–240, 243, 246, 249
- Recovery-as-a-Service (DRaaS), 101
- recovery plan, 25, 26, 44, 47, 101, 111, 167–169, 238, 239, 243, 244, 246

Distributed Control System (DCS), 35, 36, 38
Domain transformation, 8

E

Education and workforce development, 57, 221
Emergency Communication (ECO), 29, 94, 125–127, 129, 228, 235, 246–250

Encryption
- technologies, 22, 228

End-of-life (EOL), 22
Enhance organizations culture, 97
ENISA AR-in-a-BOX, 129, 232
ENISA Coordinated Disclosure of Security Vulnerabilities, 230
ENISA ECSF, 129, 231
ENISA Interoperable EU Risk-Management Systems, 129, 230
ENISA National Cybersecurity Assessment Framework (NCAF), 129, 230
ENISA Risk-Management/Risk Analysis, 129, 229
ENISA Securing Personal Data, 231
Eradication, 47, 98–100, 210

Estimate
- Data Restoration Time (EDRT), 243
- testing time, 243

EU Agency for Cybersecurity (ENISA), 77, 88, 90, 91, 127, 128, 130, 131, 133, 136, 140, 152, 154, 219, 223
EU Cyber Resilience Act (ECRA), 196, 230

European
- Cyber Crisis Liaison Organization Network (CyCLONe), 90, 223
- cybersecurity certification schemes, 131, 136

Evaluate access control, 210
Exercise of delegation, 144
Exploitation, 8, 21, 22, 31, 53, 117, 118, 153, 230
Exploratory Factor Analysis (EFA), 195
Exposed remote desktop protocol (RDP), 22

F
Factor Analysis of Information Risk (FAIR), 162
Fast Identity Online (FIDO), 209
Federal cybersecurity research and development strategic plan, 55, 56, 220
Final provisions, 81, 145–146
Firewall logs, 16
Flaws, 16, 31, 106, 116, 221, 236

G
General
　Data Protection Regulation (GDPR), 153, 231
　provision, 81–84
Generative AI (GenAI), 11, 12, 58, 61, 66
Govern, 28, 77, 98, 172–175, 178
Governance, 54, 56, 65, 77, 85, 92–93, 98, 126, 172, 173, 177–179, 187, 218, 232

H
Hackers, 42, 45, 95, 106, 107, 110, 116, 245
Holistic shift, 11
Hybrid active directory security and management, 77
Hypertext Protocol Secure (HTTPS), 121

I
Identification, 19, 22, 32, 40, 51, 52, 64, 85, 88, 96, 98–100, 103–105, 110, 113, 117, 126, 130, 167, 173, 210, 233, 235
Identify
　malicious incidents, 15
　RPO, 243
Identity governance and administration, 77
IEC 62443, 128, 137
Impact
　analysis, 237–239
Implementation Groups (IG), 184–186
Implements, 9, 23, 26, 27, 42, 44, 47, 51, 77, 78, 80, 81, 89, 91, 100, 109, 111, 113, 114, 121, 125, 128, 134, 137, 138, 155, 158, 160, 168, 177–179, 182, 184, 185, 187, 189, 201, 211, 222, 230, 232–235, 241, 247, 249
Improved security, 23
Improvements, 7, 8, 10, 14, 24, 28, 29, 33, 96, 100, 109, 110, 112, 156, 157, 159, 170, 173–175, 177, 179, 190–192, 194, 195, 197, 198, 202–211, 230, 234, 246, 247, 249
Incident
　handling, 86, 94, 98–100, 135, 227
　handling plan, 98, 99, 101
　management, 15, 79, 100, 173, 224, 234, 235
　response plan, 79, 100, 111
Indicator of Compromise (IoC), 135
Industrial
　Control and Automation System (ICAS), 34
　Internet of Things (IIoT), 2, 6, 36
Industrial Automation Control Systems (IACS), 128, 130
Industry 4.0 (I4.0)
　paradigm, 4
Information
　gathering, 116
　security, 15, 42, 43, 46, 48, 75–146, 160, 161, 184, 186, 189, 217–250)
　Security Management System (ISMS), 75, 95, 189, 191
　sharing, 81, 82, 85, 89, 91, 103, 135, 139–141, 154
Infrastructure-as-a-Service (IaaS), 5
Infringements entailing a personal data breach, 142–143
Inside
　perpetrators, 115
　test, 114, 115
Integration of Product and Process Development (IPPD), 202, 203, 208
Integrity, 20, 25, 32, 38, 39, 42–49, 55, 64, 105, 122, 123, 158, 159, 166, 169, 176, 232
Intelligent manufacturing, 6
Interconnectedness, 12, 14, 218
International Electrotechnical Commission (IEC), 75, 137, 188, 229
International Standard Organization (ISO), 75, 153, 218
Interoperability, 14, 89, 230

Intrusion Detection System (IDS), 16, 52, 65, 99
ISO/IEC
  ISO 9004:2008/2015, 192–195
  ISO/IEC 22301:2019, 129, 229
  ISO/IEC 27001:2022, 75, 109, 112, 189–191
  ISO/IEC 27002, 189, 230
  ISO/IEC 27003, 107, 189
  ISO/IEC 27004, 107, 189
  ISO/IEC 27005, 107, 159, 161, 189
  ISO/IEC 27036-2:2023, 229
  ISO/IEC 29146:2016, 129, 232
  ISO/IEC 29147:2022, 153
  ISO/IEC 30111:2019, 153
  ISO/IEC27031:2011, 129, 229
ISO TR 29156:2915, 129, 233
IT security, 14, 37, 103, 184, 231, 232, 250
IT-risks, 16, 17, 154–156, 158, 161, 163

J
Jurisdiction
  and Registration, 81, 137–139
  and territoriality, 137

L
Large Language Models (LLMs), 11, 12, 58, 60, 66
Leadership
  capability level, 202–209, 233–235
  capability maturity level, 202, 203, 205–207
  Capability Maturity Model (CMM), 201–210
  in digital transformation, 10, 11
Learning curve, 11
Likelihood, 16, 19, 20, 22, 23, 30, 40, 48–51, 79, 93, 101, 102, 109, 155, 158, 159, 161, 163, 164, 172, 173, 199, 200, 209, 236–239
Loss tolerance, 24, 26, 167, 168

M
Machine Learning (ML)
  based intrusion detection, 65
  Semi-Supervised (SSML), 62–65
  Supervised (SML), 62
  Unsupervised (UML), 62–64
Malicious
  code, 21, 22, 30, 61, 78, 103, 184, 200
  website, 53, 114

Manual mode system, 35
Mature cybersecurity posture, 23
Maximum tolerable downtime (MtD), 222, 239, 242, 244, 247, 249
Measures to Article 21.2
  basic cyber hygiene practices and cybersecurity training, 94, 113–121
  business continuity, 100–102
  human resources security, 94, 123–125
  incident handling, 98–100
  policies and procedures regarding the use of encryption, 94, 121–127
  policies and procedures to access the effectiveness of cybersecurity risk-management measures, 109–113
  policies on risk analysis and information system security, 94–98
  security in network and information systems acquisition, 105–108
  supply chain security, 102–104
  use of multi-factor authentication or continuous authentication solutions, 94, 125–127, 228
Minimum harmonization, 84
Mission-critical services, 17
Mitigating, 26, 40, 47, 49, 52, 64, 79, 105, 110, 135, 140, 163, 210
Mitigation, 16, 17, 23, 28, 40, 45, 54, 90, 99, 103–105, 107, 110, 127, 132, 139, 153–155, 158, 159, 164, 173, 174, 176, 181, 210, 220, 225, 229, 230, 234
MITRE ATTACK, 21, 104, 108, 129, 137, 151, 170, 180–183, 211, 212, 218, 230
MITRE Engage, 182–183, 212
Mobile solutions, 7
Monitoring, 6, 16, 23, 28, 32, 34, 35, 37, 44, 51, 52, 61, 64, 65, 77, 79, 95, 99, 105, 107, 109, 111, 113, 114, 127, 134, 153, 160, 164, 170, 174, 179, 188, 190, 192, 193, 207, 217, 219, 222, 224, 225, 228, 230, 244
Monitors, 6, 8, 21, 23, 28, 34, 35, 42, 46, 48–50, 52, 57, 62, 64, 77, 92, 103, 109, 113, 114, 152, 161, 164, 179, 181, 192, 209, 222
Monte Carlo Analysis, 158
Multi-Factor Authentication (MFA), 31, 94, 120, 124–127, 129, 147, 210, 218, 228, 235
Mutual assistance, 88, 89, 91, 143, 144

Index 273

N

National
  cyber crisis management frameworks, 85
  cybersecurity strategy, 82, 84, 85, 91, 153
  Initiative for Cybersecurity Education-Capability Maturity Model (NICE), 202
National Institute of Standards and Technology (NIST), 15, 55, 135, 170, 171, 220
Natural Language Processing (NLP), 3, 4, 58, 60
Network
  Disaster Recovery Plan (DRP), 167
  and Information Security (NIS2), 75–145, 171, 217–250
  traffic, 52, 63, 122, 135
  UmsuCG-OpenKRITIS
NIST
  CSF 2.0, 15, 28, 29, 75, 97, 104, 110, 112, 171, 173–180, 186, 191, 198–200, 211, 218, 229, 244
  CSF core, 56, 172
  CSF core functions, 56, 172
  CSF Organizational Profiles, 56, 172
  CSF Tiers, 56, 172, 178, 179
  Cybersecurity Framework (CSF), 15, 28, 29, 38, 55–57, 75, 84, 107, 135, 151, 170–180, 199, 220, 234
  Identify (ID), 173–175
  Detect (DE), 38, 173, 174
  Govern (GV), 173–175
  incident response lifecycle, 46–48
  Protect (PR), 38, 173, 174, 176
  Recover (RC), 38, 173, 176
  respond (RS), 173, 174
  Secure Software Development Framework (NIST SSDT), 34
  SP800-16, 129
  SP800-34, 129, 229, 233
  SP800-37, 159, 160
  SP800-45, 129, 233
  SP800-50, 129, 231, 232
  SP800-53, 113, 129, 229, 231
  SP800-55, 129, 230
  SP800-61, 129, 229
  SP800-63B, 129, 233
  SP800-100, 129, 230
  SP800-123, 230
  SP800-137, 230
  SP800-161, 104, 230
  SP800-170, 233
  SP800-171 Control, 22
  SP800-175A, 129, 231
  SP800-175B, 231
  SP800-184, 129, 229
  SP800-192, 129, 232
  SP800-209, 129, 229
  SP800-216, 129, 230
  SP1800-6, 129, 232

O

Obligations
  legal, 120
Open loop system, 35
Open Worldwide Application Security Project (OWASP)
  Comprehensive Lightweight Application Security Process (CLASP), 34
  Software Assurance Maturity Model (SAMM), 34
Operationally Critical Threat, Asset and Vulnerability evaluation (OCTACVE), 162
Operational technology (OT)
  cyber threat risk incidents, 35, 75, 219
  security (OTSEC)
Organizational Performance Management (OPM), 207
OT cybersecurity plan, 41–42
Other critical sectors, 76, 81, 218, 222, 225
OT-IT integration, 41

P

Password
  change, 81, 152
  strong, 110, 119
Patch management, 105, 111
Peer reviews, 88–91
Pegasus, 219
Penalties, 76, 77, 80, 81, 96, 143, 195, 218
Penetration
  test, 79, 95, 106–108, 116–118
Pentest-as-a-Service (PTaaS), 118, 119
Perception, 3, 16, 18, 58
Performance
  monitoring, 64
Phishing, 18, 21, 37, 45, 50, 51, 53, 61, 110, 111, 115, 121, 154, 159, 181
Physical Access Control Systems (PACSs), 34, 36
Platform-as-a-Service (PaaS), 5
Policy and Documentation Management (PADM), 30
Predator spyware, 219
Preparation, 47, 98, 100

Prepares, 27, 34, 51, 80, 89, 97, 101, 106, 131, 136, 154, 160, 166, 177, 189, 234, 237, 246, 248, 249
Preparing for NIS2, 234–236
Pre-processing, 16, 195
Prevention, 6, 24, 48, 51, 61, 94, 101, 135, 139, 141, 153, 222, 234–236
Prioritizes, 19–21, 23, 29, 31, 47, 54, 56, 97, 99, 102, 103, 110, 113, 119, 126, 141, 156, 162–165, 173, 174, 182, 211, 238
Privileged Access Management (PAM), 43, 45, 77, 210, 225
Process transformation, 8
Processing Element (PE), 59
Profile
    targeted, 175, 197
    current, 175, 177, 199–201, 204
Programmable Logic Controller (PLC), 35, 36
Protect
    sensitive data, 51, 96
Protect (PR), 38, 173, 174
Provide business continuity, 96

Q

Qualitative cybersecurity risk analysis, 48
Quality Management System (QMS), 192
Quantification, 19, 48, 103, 155, 158–162, 211
Quantum Information Science (QIS), 56, 221
Quishing, 18

R

Radar charts, 28, 29, 192, 198–200
Radio Frequency Identification (RFID), 6, 41
Ransomware, 18, 21, 45, 47, 53, 64, 89, 121, 152, 159, 177, 184, 185, 239, 245
Reconnaissance, 116
Recover, 17, 25, 27, 29, 38, 43, 47, 52, 54, 77, 79, 101–103, 114, 139, 166, 167, 173, 174, 176, 178, 199, 217, 233, 237, 243, 244
Recovery
    Point Objective (RPO), 24, 25, 52, 101, 102, 111, 165, 166, 222, 237, 239–242
    Time Objective (RTO), 24, 25, 52, 101, 102, 165, 166, 222, 237, 239–244
Reduce costs, 97, 202, 204
Registry of entities, 138
Regular updates and adaption, 112
Regulatory compliance, 23, 38, 52, 96, 111, 152, 196
Reporting compliance, 52
Report on the state of cybersecurity in the Union, 90
Reputation, 12, 14, 22, 37, 45, 46, 48, 94, 99, 102, 103, 110, 113, 126, 165, 166, 200, 244
Resilience of Critical Entities (RCE), 217–219
Respond, 15, 21, 23–25, 28, 29, 33, 34, 38, 48, 50, 55, 60, 64, 77–79, 100, 101, 104, 106, 109, 113, 120, 135, 139, 164–167, 171, 173, 174, 176, 178, 179, 196, 199, 202, 209, 220, 222, 227
Responses, 4, 5, 17, 20, 21, 23, 28, 39, 46–49, 52, 54, 60, 63, 79, 82, 85, 86, 89, 90, 93, 98–104, 111–113, 128, 130–132, 135, 139, 140, 146, 153, 165, 173, 174, 176, 178, 183, 187, 188, 195, 196, 210, 211, 224, 225, 227, 234, 238–240, 244, 245, 247, 250
Review, 33, 42, 54, 77, 88, 89, 91, 102, 105, 109, 118, 130, 145, 163, 167, 182, 190–196, 210, 218, 219, 232, 241
Review security awareness and training, 210
Risk
    analysis, 33, 48, 49, 94–98, 129, 154, 226, 229, 235
    assessment, 15, 16, 19–21, 28, 48, 77, 88, 91, 96, 97, 108–110, 113, 128, 130–131, 137, 155, 156, 161–163, 174, 175, 178, 190, 198, 210, 223, 225, 238, 247
    documentation, 154
    evaluation, 20–21
    identification, 19, 130, 154, 156–158
    level, 20, 21, 48–50, 159, 160, 237
    management, 13, 79, 151–170, 217
    review, 54, 218
Root cause analysis
    management objectives, 98, 175
    management requirements in OT systems, 38
    mitigation, 16, 107, 110, 154, 155, 159, 210, 220, 229, 234
    quantification, 19, 48, 155, 158–162
    tolerance, 16, 46, 98, 172, 175, 179, 184

S

Sanctions, 76, 77, 80, 81, 96, 114, 223
Scopes, 9, 17, 27, 47, 60, 76, 77, 82, 84, 95, 99, 108, 115, 116, 130, 137, 139, 141, 158, 162, 167–169, 177, 189, 190, 193, 195, 208–210, 217, 226, 234, 239

Sectors of high criticality, 76, 81, 218
Sector-specific union acts, 82
Secure
    Contain Protect (SCP), 121
    by Design and Default (SbD), 53, 221
    Shell or Socket shell (SSH), 122
    Socket Layer (SSL), 121
    Software Development Lifecycle (SDLC), 33
Security
    in development, 236
    Event Management (SEM), 52, 194
    Information and Event Management Systems (SIEM), 43, 52, 111
    Operations Centers (SOCs), 54, 76, 90, 135, 181, 218
Self-assessment, 88, 191–193, 200
Single-Sign-On (SSO), 31, 94, 126
Skilled employees, 2, 206
Smishing, 18
Social Engineering
    attack, 53, 61, 211
    audit, 53, 61, 114, 115, 211
Software
    as-a-Service (SaaS), 5
    Bill of Material (SBOM), 127
    Engineering Institute (SEI), 208
Spear phishing, 53, 181
Spider diagrams, 28, 192
Spyware, 219
Standardization, 75, 136–137
Standard Query Language (SQL), 103, 122
Subject matter, 82, 171
Supervision and Enforcement, 81, 141–144, 195, 223
Supervisory
    and enforcement measures in relation to essential entities, 142
    and enforcement measures in relation to important entities, 142
    Control And Data Acquisition (SCADA), 35
Supply chain
    security management, 236
SWOT analysis, 156, 157, 162
Systems Security Engineering Capability Maturity Model (SSE-CMM), 202, 203

T
Tactics, Techniques and Procedures (TTPs), 50, 181, 182
Technology
    emerging, 10
Third-party risk
    management, 111
Threat Agent Risk Assessment (TARA), 163
Timeliness and performance requirements, 38
Time To Detect (TTD), 26, 168
Transport Layer Security (TLS), 122
Transposition, 87, 145, 234
Trustworthy Distributed Digital Infrastructure (TDDI), 57, 221

U
Unified endpoint management, 78
Unpatched software, 22, 38, 95

V
Virtual simulation models, 6
Vishing, 18
Voluntary notification or relevant information, 140–141, 235
Vulnerability
    assessments, 79, 95, 218
    disclosure, 31, 33, 86, 88, 105, 106, 153, 154, 230
    Disclosure Policy (VDP), 107
    management, 40, 105, 230, 236
    report, 31, 79, 107
    testing, 106
    tracking, 106
Vulnerable operations systems, 22

W
Weaknesses, 18, 19, 23, 29, 31, 32, 40, 50, 52, 79, 95, 102, 115–118, 156, 162, 189, 192, 199, 201, 209, 211, 237
Wi-Fi, 117
Withstand, 43, 53, 116, 186
Workforce efficiency, 9

Z
Zero-trust, 57, 124, 129, 154, 227, 250

MIX
Papier aus verantwortungsvollen Quellen
Paper from responsible sources
FSC® C105338

If you have any concerns about our products,
you can contact us on
ProductSafety@springernature.com

In case Publisher is established outside the EU,
the EU authorized representative is:
**Springer Nature Customer Service Center GmbH
Europaplatz 3, 69115 Heidelberg, Germany**

Printed by Libri Plureos GmbH
in Hamburg, Germany